NICK HERBERT has a doctorate in physics from Stanford University and is the author of two previous books, *Quantum Reality* and *Faster Than Light*. He has directed physics seminars and internal conferences on quantum physics at the Esalen Institute in California. He lives in Boulder Creek, California.

ALSO BY NICK HERBERT

Quantum Reality: Beyond the New Physics
Faster Than Light: Superluminal Loopholes in Physics

NICK HERBERT

elemental
mind

Human Consciousness and the New Physics

A PLUME BOOK

PLUME
Published by the Penguin Group
Penguin Books USA Inc., 375 Hudson Street, New York, New York 10014, U.S.A.
Penguin Books Ltd, 27 Wrights Lane, London W8 5TZ, England
Penguin Books Australia Ltd, Ringwood, Victoria, Australia
Penguin Books Canada Ltd, 10 Alcorn Avenue, Toronto, Ontario, Canada M4V 3B2
Penguin Books (N.Z.) Ltd, 182–190 Wairau Road, Auckland 10, New Zealand

Penguin Books Ltd, Registered Offices: Harmondsworth, Middlesex, England

Published by Plume, an imprint of Dutton Signet,
a division of Penguin Books USA Inc.
Previously published in a Dutton edition.

First Plume Printing, November, 1994
10 9 8 7 6 5 4 3 2 1

 REGISTERED TRADEMARK—MARCA REGISTRADA

The Library of Congress has catalogued the Dutton edition as follows:
Herbert, Nick.
    Elemental mind : human consciousness and the new physics / Nick
Herbert.
        p. cm.
    Includes bibliographical references and index.
    ISBN 0-525-93506-1
            0-452-27245-9 (pbk.)
    1. Consciousness. 2. Quantum theory. I. Title.
B808.9.H47   1993
128'.2—dc20   93-15700
                                        CIP

Printed in the United States of America
Original hardcover design by Steven N. Stathakis

To Dear Old Mom

contents

acknowledgments

Physics, in which I am well trained, is the science of matter. Mind, on the other hand, is something else entirely: the topic of this book. What's a guy like you, Nick, doing in a place like this?

I'd like to thank my teachers in physics who taught me what I know about matter, including Philip S. Jastram, Leonard Jossem, Bob McAllister (my welding instructor at Ohio State), Walter Meyerhof, Leonard Schiff, Wolfgang Panofsky, Felix Bloch, Sidney Drell, and many others.

I was dramatically introduced, in 1963, to consciousness as a research problem by my psychology colleagues Elizabeth Rae Larson and Ann Manly. Doctors Paul Rosenberg, Robert

Erickson, and Bill Ross, and Tom Records, Janice Blue, Allison Kennedy, Harry Eli, Elizabeth Gay, Betsy Rasumny Riñen, Philippa Meyering, and Dave Whittaker were also instrumental in guiding me toward a better appreciation of the mystery of mind.

I would like to thank Arthur Young of the Institute for the Study of Consciousness in Berkeley for hosting many meetings of our Consciousness Theory Group and the members of that group for many happy evenings exploring our favorite topic. Thanks to Saul-Paul Sirag, Jerry White, John Holmdahl, Elizabeth Rauscher, Barbara Honegger, Michael Rossman, Jean Burns, Patricia Rife, Ruth-Inge Heinze, Jon Klimo, Jack Engstrom, Jack Karush, and Michael Karnov. Henry Dakin also deserves recognition for supporting the CTG in San Francisco, and for helping me produce an earlier version of this book.

I would like to thank Mike Murphy and the late Dick Price for opening up Esalen Institute, Big Sur, for many years to invitational seminars on the physics of consciousness. I would like to thank the many participants in these seminars, including Saul-Paul Sirag, Henry Stapp, John Clauser, Philippe Eberhard, John Cramer, Bernard D'Espagnat, Dieter Zeh, Ariadna Chernovska, Jeffrey Bub, Itamar Pitowski, Larry Bartell, Richard Baker Roshi, Beverly Rubik, Rudi Rucker, Ralph Abraham, Tom Etter, Dana Massie, Gene Bernard, Charles MacDermed, Lila Gatlin, David Finkelstein, Beverley Kane, Mary-Minn Peet, Fred Wolf, and Jack Sarfatti. I would like to thank the many members of the Esalen staff who mingled their views with ours, including Nancy Lunney, Jane Miletich, Diane Miller, Al Drucker, Stan Grof, Joan Halifax, George Leonard, and Gregory Bateson. I would also like to acknowledge support by Werner Erhard and by George Koopman of my consciousness studies.

I thank Charles Brandon, for discussions on transcendence and for supporting the Reality Prize awarded at Esalen in 1982 to John Bell and John Clauser to honor their experimental proof of nature's basic nonlocality.

For opening my mind to wider dimensions of consciousness I'd like to thank members of the FOG, Isthmus, Island, and International Synergy groups including Ralph Abraham, Andra Ackers, Larry Dossey, David Dunn, Will McWhinney (Captain Ambiguity), Alan Brodsky, Paul Lee, Terence McKenna, Ralph Metzner, Rupert Sheldrake, Jill Purce, Bruce Eisner, Nina Graboi, Peter Stafford, and Elizabeth Gips.

For convening the Noetics Institute seminars on New Models of Life, I'd like to thank Beverly Rubik and the participants, including Geoffrey Chew, Henry Stapp, Amit Goswami, Paul Lieber, Dick Strohman, Harry Rubin, Willis Harman, George Weissman, Dick Blasband, Ted Roszak, and others.

For hospitality on the road, I'd like to thank Tanis Fletcher, Peter and Ida Scott, Brian Wallace and Faustin Bray, Larry and Mari Thorpe, and Allison Kennedy and Ken Goffman.

For patiently explaining to me their theories of mind and for good fellowship, I'd like to thank Saul-Paul Sirag, Evan Harris Walker, Harry Klopf, Doug Seeley, and James Culbertson.

For much discussion about reality issues and the problems of consciousness, I thank my friend the late Heinz Pagels. I miss you, Heinz.

Thanks to my tough-minded agent, John Brockman, my patient editor, Rachel Klayman, and to many others not listed here whose presence has changed my life.

Thanks to my wife, Betsy, and son, Khola, for tolerating my absentmindedness while completing this book.

introduction

Why subjective experiences at all? Why is it that I experience anything?
Why don't I just go ahead and do what I'm doing without any experience?
—STAN FRANKLIN

Is there a mind/body problem? And if so, which is it better to have?
—WOODY ALLEN

Nothing in nature is more mysterious than the human mind.
Where does it come from? What makes it work? And where
does the mind go when we die? Philosophers and religious
thinkers have offered a bewildering variety of answers to
these questions, but only recently has science begun to tackle
the problem of human consciousness.

In the past 50 years, the frontiers of physics have advanced from middle-range phenomena to the large-scale problems of cosmology and the small-scale physics of elementary
particles. In principle, physics on the scale of everyday life is
completely understood. At this stage of maturity, there is no
excuse save lack of imagination for physical science not to attempt to provide a technical solution to the mind/body problem

(in place of the merely verbal solutions proposed by philosophers). We might expect the technology arising from a scientific understanding of the mind to create radically new mental experiences, novel modes of being, artificial forms of consciousness, as well as eliminating our utter ignorance concerning the true place of human minds in the community of sentient beings. Whatever progress science makes in its study of mind, I believe that we cannot say that we really understand consciousness until we can actually build things that have inner experiences like our own.

So in 1974 I tried to construct the world's first conscious machine.

It didn't work.

The unsuccessful conscious machine (called the "metaphase typewriter") was one project of the Berkeley-based Consciousness Theory Group, whose goal was a no-holds-barred investigation of the origins of inner life in humans and other sentient beings. To this task we brought expertise in physics, neuroanatomy, computer science, philosophy, and theology.

The core of the Consciousness Theory Group consisted of me, Saul-Paul Sirag, John Holmdahl, and Jerry White.

Trained as an experimentalist at Stanford, I worked in the San Francisco Bay Area as an industrial physicist developing magnetic, optical, and electrostatic data storage devices. Besides the mystery of consciousness, my main preoccupation has been with the foundations of quantum theory, especially Bell's connectedness theorem.

Saul-Paul Sirag, born in Borneo of Dutch-American missionary parents, is a self-educated physicist specializing in multidimensional models of matter and mind.

John Holmdahl, son of a California senator, was our guide to the neuroanatomy of the brain; Jerry White is a philosopher and computer scientist with a talent for learning obscure languages such as COBOL, Mongolian, and ancient Akkadian.

The Consciousness Theory Group originally met in Berkeley at Bell helicopter designer Arthur Young's Institute for the Study of Consciousness. We also shared spacetime events with the Berkeley Brain Center (BBC) group centered around Fred Lenherr, Richard Hodges, and Elaine Chernoff. In the early seventies, Berkeley was a ferment of well-educated people passionately seeking the secret of ordinary awareness. In those days, it seemed that one could not visit the Safeway without running into a mind scientist or two in the produce section.

The Consciousness Theory Group later moved across the Bay to San Francisco, setting up shop in toy manufacturer Henry Dakin's Washington Research Center, where an early version of this book was written. At about this time, Michael Murphy invited members of the group to gather at Esalen Institute in Big Sur. At one of these Esalen meetings, our group, in conjunction with Charles Brandon's Reality Foundation, awarded the first Reality Prize to physicists John Stewart Bell and John Clauser for their discovery of nature's essential quantum interconnectedness. To these people, and many more unmentioned, I am grateful for companionship, inspiration, and support.

Two major conjectures dominate the scientific debate on the nature of mind: (1) mind is an "emergent feature" of certain complex biological systems; (2) mind is the "software" controlling the brain's computerlike hardware. *Elemental Mind* explores a third hypothesis—that, far from being a rare occurrence in complex biological or computational systems, mind is a fundamental process in its own right, as widespread and deeply embedded in nature as light or electricity.

Along with the more familiar elementary particles and forces that science has identified as building blocks of the physical world, mind (in this view) must be considered an equally basic constituent of the natural world. Mind is, in a word, elemental, and it interacts with matter at an equally elemental level, at the level of the emergence into actuality of individual quantum events. The behavior of matter at the quantum level affords both the opportunity for mind to manifest itself in the material world and the means for us to explore the details of the mind's operations "from the outside," as it were, in addition to the private access to mind "from the inside" that we enjoy in common with other sentient beings. In this view quantum theory offers a royal road to a new science of mind.

Three features of quantum theory are especially suggestive for understanding how mind might enter matter at the quantum level. Coincidentally, these three features—randomness, thinglessness, and interconnectedness—were precisely the features that Albert Einstein, one of quantum theory's founding fathers, found so bizarre that he could not accept them. These three Einstein-abhorred features, however, have continued to play an important role in quantum thinking; quantum connectedness in particular has been securely confirmed by recent experiments motivated by the theorem of Irish physicist John Bell. *Elemental Mind* makes a plausible case from biological, psychological, and parapsychological evidence that these three features of matter are the external signs of three basic features of mind: free will, essential ambiguity, and deep psychic connectedness.

One of the major mistakes of the medieval philosophers was their underestimation of the size of the physical world. This cozy earth, the seven celestial spheres, plus Dante's concentric circles of hell: that was the full extent of the universe in the medieval imagination. No one at that time even dreamed of other solar systems, let alone galaxies like dust in a vast room billions of light-years in diameter.

I believe that modern mind scientists are making this same medieval mistake by vastly underestimating the quantity of consciousness in the universe. If mind is a fundamental force in nature, we might someday realize that the quality and quantity of sentient life inhabiting just this room may exceed the physical splendor of the entire universe of matter.

James Watson, the codiscoverer, along with Francis Crick, of the spiral structure of DNA, once remarked: "I don't think that consciousness will turn out to be something grand. People said there was something grand down in the cellar that gave us heredity. It turned out to be pretty simple—DNA."

I confess that I do think that consciousness will turn out to be something grand—grander than our most extravagant dreams. I propose here a kind of "quantum animism" in which mind permeates the world at every level. I propose that consciousness is a fundamental force that enters into necessary cooperation with matter to bring about the fine details of our everyday world. I propose, in fact, that mind is elemental, my dear Watson.

steps toward a science of consciousness

And never for each other shall we feel
As we may feel, till we have sympathy
With nature in her forms inanimate,
With objects such as have no power to hold
Articulate language.
In all forms of things there is a mind.
                                        —WORDSWORTH

The mind alone sees and hears; all else is deaf and blind.
                                        —PLATO

"I'm afraid, Dave!"

When the spaceship's computer in Stanley Kubrick's film *2001* has its brain removed, we sympathize more with Hal the frightened machine than with the human astronauts it has murdered. What is machine consciousness that men should be so mindful of it?

In 1950, computer pioneer Alan Turing devised the famous "Turing test" for machine intelligence. A computer in a box passes the Turing test if it can convince a human being that there is another human in the box. Simple dialog programs such as Joseph Weizenbaum's ELIZA or Bill Chamberlin and Tom Etter's RACTER already carry on plausible conversations that, in some cases, reduce their human partners to tears. (Real machines don't cry.) For me the Turing test misses the point: it seems highly unintelligent to base the important question of whether a machine is conscious or not on human gullibility.

I sometimes imagine myself as a language coach at a school for robots, preparing them to pass their annual Turing tests. "Humans are particularly sly and devious animals," I say, "and they appreciate a certain amount of deviousness in their electric competitors. Try to avoid the typically robotic temptation to give a direct answer to a direct question. For instance, if a human asks: 'Come on now, level with me. Are you really conscious or not?' a good reply might be: 'Don't be silly. Heh-heh. A machine can't think.'" The novice robots and I spend the rest of the afternoon practicing sly "heh-heh"s while the advanced robots learn the art of evasive behavior by watching human expert systems make political speeches on TV.

Since humans are a fairly credulous lot, we shouldn't be impressed by how many folks the Hal 9000 can fool with its frightened-human-being routine. What I really want to know is not how good a show of emotion Hal can put on, but whether he actually "feels" the emotions that his behavior seems to express.

Consciousness—What's the Problem?

Consciousness, at least as humans experience it, has little to do with performance. Much of what we do—and do exceedingly well—is better done outside conscious awareness. Without the slightest conscious effort I digest my potatoes, beat my heart, and defend my body against hostile bacteria. Even activities for which consciousness seems to be essential, such as learning a new piano piece, gradually become automatic upon repetition, requiring less and less attention for their skillful execution. Consciousness seems not to be concerned so much with what an entity does as with what it experiences while it is doing it. What we would like to know about Hal, or for that matter, about any other alleged conscious beings from our spouses to pet cats, is whether he possesses "insides" like ours, or whether he is merely an empty automaton, just "going through the motions."

To put the matter bluntly, as one center of sentience to another, conscious beings like us have "insides" (experiences) as well as "outsides" (behavior). Unconscious beings—including us during interludes of deep sleep or coma—possess only outsides. No matter how complex its behavior, a being with no insides might well be called a "mere thing." Good manners suggest that, no matter how they behave, creatures with insides should be addressed as "persons."

To appreciate the depth of the conceptual gulf that separates consciousness from behavior (insides versus outsides) imagine a creature that possesses a rich inner life but exhibits no behavior at all, for instance, the paraplegic war veteran in Dalton Trumbo's novel *Johnny Got His Gun*, trapped inside a body that does not respond to his will. Cal Poly consciousness theorist James T. Culbertson calls such examples of helpless sentience "paralyzed conscious robots." A paralyzed conscious robot has lots of insides but hardly any outsides at all. If you push such a robot, it falls over—a type of response it shares with all other inanimate objects—but this simple ballistic behavior gives no clue whatsoever to the presence of an ongoing conscious experience inside the tumbling robot.

More complex behavior than the act of falling off a shelf could suggest the presence of inner life, but it seems to me that no conceivable robotic behavior could ever prove beyond a doubt that the robot actually possessed insides. For the original Turing test, Alan Turing envisioned a simple teletype as the alleged conscious computer's input/output device. Suppose we expand the computer's repertoire of inputs to include sight, hearing, touch, as well as taste and smell. Let's give the machine a pleasant synthetic voice and a humanoid body with a full range of expressive gestures (including the ability to shed tears) and cover the whole thing with warm and responsive artificial flesh. Certainly if the original Turing teletype machine could convince certain gullible humans that there was a human being inside the box, then a Turing humanoid, with its wider range of ways to simulate the expression of human feelings, could seduce even more folks into thinking that the talking doll with the polyurethane skin was actually having inner experiences.

The question of pretty robots aside, how do we know that other human beings are conscious? Like the situation with the Turing teletype or humanoid, the only clue that we have to the presence or absence of consciousness in another human being is how that being behaves. I know that I myself am conscious via an undeniably direct and immediate revelation. But what about my neighbor? Could it be that he is just going through the right motions but there is actually nobody at home? Philosophers call the question of how to decide whether your neighbor is a soulless robot or a sentient being "the problem of other minds."

The philosopher's other-mind problem is complicated by what I call "the Grand Illusion": the persuasive conviction that the entire universe is centered around my self. (You probably suffer from a variant of this illusion: the belief that the world revolves around *you*.) When I look at my own experience I do indeed appear to be located at the center of the world, a bright focal self full of intense sensations and feelings, compared with which the rest of the world seems drab and devoid of feeling. If other such centers of intense sentience exist, as suggested by indirect evidence, then the Grand Illusion must be discarded as a kind of mirage, a seductive but ultimately unreliable guide to reality.

The Grand Illusion resembles a certain optical illusion called in German *Heiligenschein*, or "holy halo." On a cold sunny morning, look at your shadow in the dew-covered grass. Your head will seem to be surrounded by a blaze of white light. But the shadows of your companions do not glow; only your own head seems to be blessed with the holy halo. Famous sixteenth-century Italian artist Benvenuto Cellini viewing his solo halo in the grass took this phenomenon as a sign of his own genius. The light shines only around my head, not around anybody else's. Likewise my direct experience of my own awareness, contrasted to my very indirect appreciation of other people's experiences, not to mention nonhuman forms of experience, tends to make me feel alone in the world, isolated in solo reverie.

Because our experience of our selves is so intense, our experience of others so weak by comparison, many of us have at least flirted with accepting the Grand Illusion as plain fact, believing, at least for a time, that only one conscious being exists in the world, namely me, and all other creatures are soulless zombies. Philosophers call this (presumed) illusion *solipsism*. The few people foolish enough to act out this belief are labeled "sociopaths" by the criminal justice system, which punishes them for their totally self-centered behavior no matter how strong the philosophical arguments they might muster for their pontifical position. Since a society of solipsists is a contradiction in terms (there can be only one solipsist in the universe), all societies necessarily reject the Grand Illusion as a guide to human conduct. Social conventions aside, though, how can an individual like me logically escape the solipsistic fallacy and establish to my own satisfaction the existence of other minds? Bertrand Russell once said that solipsism is completely irrefutable but boring: we should just ignore solipsism in favor of less defensible but more interesting models of mind. But the existence of other minds should be based on better criteria than escape from boredom. In particular can the guy next door pass the Turing test by performing some public act that only conscious beings are able to do? Can he actually do something to prove to me that he has insides like my own?

In the first half of the twentieth century American psychology was dominated by the behaviorist movement. The behaviorists rejected traditional introspective psychology as sterile and unscientific, as little more than a disconnected collection of stories and anecdotes, more literature than science. Both as an antidote to introspective vagueness and in the style of the so-called hard sciences of physics and chemistry, the behaviorists proposed to create the world's first truly scientific psychology out of external data alone, without resorting to unreliable subjective reports.

The behaviorists believed, at least as a working hypothesis, that whatever might go on "inside" an organism was irrelevant to a scientific explanation of that organism's behavior. They proposed to treat all organisms, including humans, as black boxes, hoping to discover objective laws relating the box's inputs (stimulus) to the box's behavior (response) without ever having to include the box's "experiences" as a factor in their calculations. Since it ignores what seems to be the most important feature of human life—namely what it feels like from the inside—the behaviorist approach to human psychology seems doomed from the outset. One could imagine that certain automatic reflexes might be handled by this simplistic approach, but whole ranges of complex human behavior would simply remain incomprehensible to behaviorists. In particular, an easy way to confound the entire behaviorist enterprise would be to point out a single example of a type of behavior for which the stimulus/response model fails, a type of behavior that is "consciousness-specific," that is, a form of outer expression that cannot be explained without taking the organism's inner experience into account.

Traditional introspective psychologists were highly motivated to overturn the behaviorist program by discovering some sort of consciousness-essential behavior. The introspectionist's goal was a sort of Turing test in reverse, in which the introspection test is given to humans not robots. Whereas the Turing test asks for some sort of robotic behavior that will convince a person that the robot is a kind of human being, the antibehaviorists sought a kind of human action that would convincingly show that a human being is more than a robot. Behaviorism is now passé, having been replaced by other fashions in psychology: awareness- and body-centered therapies such as gestalt and bioenergetics, and a fascination with altered states of consciousness such as dreaming, meditation, and hypnosis.

Although psychology has returned to a more "humanistic" orientation, it is important to realize that behaviorism was never refuted. In particular, no enterprising antibehaviorist was able to come up with a type of behavior for whose explanation consciousness was essential. Psychologists did not defeat behaviorism but merely moved on to wider concerns than stimulus/response research. Although they failed in their attempt to bring all psychology into their camp, the behaviorists succeeded in calling attention to a crucial weakness in experimental psychology. In fact, for the science of consciousness behaviorism's most enduring legacy might be this: we now know that the tools of twentieth-century science are powerless to verify the presence of consciousness in human beings—the one system in the universe that we know with certainty possesses it. The behaviorists in effect issued a challenge to modern experimental science to come up with an objective way to measure the presence of subjective experience. So far science has utterly failed to meet this challenge. The fact that certain subjective states such as dreaming or meditation are correlated with particular patterns of electrical activity recorded from electrodes attached to the scalp (so-called brain waves) is no more an objective indication of the presence of consciousness than a sensible conversation carried out by ELIZA. Suppose an electronic device (such as the tape recorder that stored the brain-wave signals) produced the same pattern of electrical signals as a dream-state electroencephalogram (EEG). Would we conclude from these signals that the tape recorder was dreaming?

An important side effect of our inability to measure the presence of consciousness is that there is no scientific way to verify that a system is *unconscious*. Thus the commonsense belief that stars, rocks, and atoms are unconscious has no real scientific basis and should rightly be regarded as groundless superstition. The belief that matter is "dead" has the same experimental status as the opposite animistic belief that matter is "alive." Both beliefs rest on an equal logical footing, although the animist can in his favor point to at least one material system that is "alive" while the materialist cannot point to any kind of matter that he knows with certainty is "dead." The real status of the inner life of "inanimate" objects awaits for its resolution a deeper kind of science than we currently possess.

The major barrier to the development of a true science of consciousness is our lack of any objective way to tell whether a given chunk of matter is conscious or not. At present our only sure means of assessing the inner life of (certain forms of) matter is by introspection and by inference. I know—without any doubt: Cogito, ergo sum—that I am conscious via immediate revelation, a direct insight compared to which all other forms of knowledge are secondhand and indirect. If present-day science finds itself powerless to validate my private insight into the real nature of things, so much the worse for science. Second, I surmise, not entirely certain, that you are conscious too, because (1) you behave like me and (2) you have a similar (biological) origin. Your outsides and your history are much like my own, so that I imagine that your insides are similar too. But computers, or even handsome soft-skinned robots, no matter how well behaved, have a life story radically different from my own. It's a big jump to infer from its behavior that a robot is conscious, while a human being acting exactly the same would surely be judged to be acting from an experiencing center.

Experiments to Detect Inner Experience

Although we presently possess no objective test for the presence of awareness in matter, the possibility of devising methods for the detection of subjective states does not appear to be entirely unthinkable. I can imagine at least three ways that subjective states could become accessible someday to scientific scrutiny.

The "Purple Glow Effect"

Suppose there is a small central portion of the brain that is responsible for consciousness while the rest of the brain is a mere unconscious computer harnessed to the service of the brain's central authority. (As evidence for the notion that consciousness is a localized brain function, we observe that large portions of the cortex and other segments of the central nervous system can be removed without loss of consciousness.) Suppose moreover that whenever a person is awake (not asleep, comatose, or "absentminded") this crucial portion of the brain emits a distinctive purple glow. Unlike electrical signals, which are produced at all times, this glow occurs only in association with conscious experience. Suppose that further research shows that although the purple glow is physically identical to ordinary light, its production cannot be explained by normal electromagnetic mechanisms. In order to fit the purple glow phenomenon into physics, scientists must invoke a previously unsuspected new force linking light and the inner life, a link that mystics have metaphorically celebrated for centuries. (It goes without saying that experimental evidence for purple glow or any other special physical manifestation of awareness is nil.)

Alternatively, instead of ordinary light, conscious entities might signal their presence by emitting new kinds of elementary particles (cogitons?). Cogitons might be detectable by physical means or they might interact only with other conscious beings, leading to a new class of particle detector. Thus the human mind would act as a sort of psychic Geiger counter.

The "purple glow," "cogiton," or similar phenomena would represent a simple and direct answer to the behaviorist challenge. These phenomena are (or would be, if they existed) types of physical behavior uniquely associated with consciousness and with no other physical process. The presence of such phenomena would open consciousness (human and otherwise) to examination via the same objective methods that have been so successful in the physical sciences. Purple-glow physicists might even detect the purple glow in inanimate systems and draw the conclusion (scientifically based, not mere prejudice) that stars, rocks, and atoms are conscious.

Mental Telepathy

It's possible that consciousness produces no unique physical signature but that its presence can be detected by certain human beings who might be called "empaths."

Empaths may directly feel another organism's state of consciousness as a certain movement inside their own minds. Or they may sense the presence of awareness in other beings indirectly as an "aura" or "chi flow," a certain visual impression invisible to ordinary people. The aura may not objectively exist in the sense of being detectable by suitably placed physical instruments. Instead these private signs of the presence of consciousness may be self-induced modifications of the empath's visual field, resembling somewhat the studio-induced captions on a live TV image, a co-option of the empath's visual field for the presentation of a nonvisual type of information.

Since this kind of consciousness detection is indirect and depends on the subjective report of another human being, one may be reasonably skeptical of an empath's reports. However, confidence in the empathic method would increase if several independent empaths agreed among themselves and with EEG measurements concerning the exact moment when a target patient recovered from a state of general anesthesia.

One can imagine empaths sensitive only to human forms of awareness who can sense a paralyzed person's call for help or ei⬤aths who have expanded the range of their "sixth sense" to embrace animals, or even computer mainframes. Alan Turing himself recognized the weakness of his famous Turing test as a reliable diagnostic instrument for consciousness, and, somewhat in desperation, suggested that if a computer could exhibit extrasensory perception, then we might believe that it possessed an inner life.

One of the major drawbacks to the use of human empaths as consciousness detectors is that they would probably not be sensitive to nonhuman forms of mind. For instance, a being's internal pace of events, what might be called its "inner tempo," must be quite different for the mind of a giant sequoia, a snowshoe rabbit, or a salmonella bacterium. If we plan to assay the awareness of our computers with mental telepathy, we will face a problem similar to that of talking with divers whose speech frequencies have been speeded up by their breathing mix. We will need at the very least a method of pacing the inner tempo of human minds to the corresponding tempo of their nonhuman sentient partners. However, once inner rates are matched between beings, there remains the problem of making sense of a completely alien set of inner experiences. Effective telepathic contact between one inside and another would most likely be restricted to human/human links for a long time to come.

Mind Links

The most direct and convincing proof that another human being, animal, or computer is conscious would be actually experiencing for yourself what that other being is feeling, with a quality and intensity comparable to your own self-consciousness. The sharing of another person's inner life, not by inference, empathy, or analogy but by merging of the two insides into a new type of co-conscious experience, would certainly constitute powerful evidence for the presence of inner experience in that other being, no matter what that being's outward behavior might be.

To determine whether your newly constructed robot is conscious or not, connect your self directly to the machine's "interior" with a "mind link." If the machine is "alive," you will experience an augmentation of your familiar human style of awareness by the robot's distinctive mechanistic form of inner life. If the machine is a "mere thing" (or if the mind link is inoperative), the linkage with the robot will not change your conscious experience.

In the case of robot/human co-consciousness, I envision the mind link as some sort of material connection between brains (or between brains and central processing units), a new kind of communication channel that transmits more than feeling-free data: the patterns of another being's inner life flow down the mind link's vital pathways. Unlike telepathy, which presumably operates in some mysterious mental realm accessible (if at all) only to a few skilled empaths, the mind link would be a purely physical connection, open to everyone, as public as the telephone. The mind link's invention would solve the problem of other minds in the simplest possible way, by making the presence and contents of other minds publicly available in a manner as direct and undeniable as the presence and contents of your own mind.

The Searle Test for Artificial Awareness

Philosopher John Searle at the University of California at Berkeley holds that "the only way to tell if a physical system is conscious or not is to be that system." John knows for sure that the Searle brain is conscious because he directly experiences that brain's inner life. The Searle test for inner life seems at first glance to be just another recipe for solipsism: "I'm conscious; I don't know about you." But Searle has come up with at least one way to extend his awareness test to forms of matter other than the brain he was born with. Suppose we want to know whether silicon computer chips can support a style of inner experience like our own. Searle proposes a test for silicon-based awareness that involves replacing his brain's neurons one-by-one with silicon chips that perform the same function. (Searle tacitly assumes that we know what the "function" of a neuron is.) At the end of this replacement process, John's skull, once the containment vessel of a meat brain, is now completely filled with computer chips. This new computer passes the Searle test for consciousness not if John merely says that he is self-aware, but only if he truly feels that he is conscious.

At a recent conference on the scientific study of consciousness, Searle described three conceivable outcomes of his replacement test for silicon-based awareness. First, the test could succeed: his new chip brain would produce behavior equivalent to the external activity produced by the old Searle meat brain. More important, the chip brain also would produce an inner life indistinguishable in quantity and quality from his old experiences.

Second, the operation could leave John totally paralyzed, devoid of any behavior but possessed of a normal flow of inner experience. He would hear the doctors expressing regret over the apparent death of the Searle brain but the inner Searle could not tell them that he was still alive.

Third, as more and more of Searle's meat brain was removed, he would feel a gradual diminution of his inner life but his body would persist in its usual behavior. As the final chips were being exchanged for the last neuronal circuit, Searle would be dimly aware of a familiar voice, seeming to come from a great distance, saying to the doctors: "Yes, yes. I feel fine. The operation was a success." Then all awareness would cease: John Searle would be dead. An "empty" Searle zombie would get up from the operating table. In this third case, after it returned to the university, the Searle zombie would find itself in the unusual position of unconsciously giving lectures on the subject of consciousness, certainly not the first time a university professor gave a lecture on a subject of which he had no firsthand knowledge.

Every science has its own circumscribed subject matter, its body of experimental facts, and its array of theories to explain those facts. The new science of consciousness will also have its proper scope, its crucial experiments, and its explanatory theories. The scope of the fledgling science of consciousness is the inner life of human (and nonhuman) beings. The difficulty of performing experiments on the inner lives of beings other than myself is one of the main barriers to establishing a firm factual basis for a science of inner life. About the inner life of nonhuman animals we know very little; about the inner life of nonbiological beings we know absolutely nothing.

Imagine being a worm in the ocean that possessed only one external sense—the sense of hearing. What kind of a physics could such a being create from its varied auditory experiences? No astronomy, no optics, no theories of electrical or magnetic phenomena. No chemistry, mechanics, or geology. Probably the best theory of the world that this single-sensed being could muster would be a musical insight into the nature of vibrations. Like those of this ear-logical marine worm, our experiences are limited to only one kind of awareness—the human kind—in only one body, with only indirect access to that same kind of awareness in other bodies. Our limited access to other minds severely restricts the kind of facts we can collect about the variety of inner experience that might exist in the world. As the poet William Blake sang: "Who knows but every bird that cuts the airy way, Is an immense world of delight, closed to our senses five?"

Although we have direct access to only one style of inner life, the quality and quantity of human awareness vary widely even in the course of normal life: from stupor to the heat of intellectual passion; from sleep to sexual ecstasy. Dissatisfied with the usual variations on the theme of ordinary life, many men and women have pursued methods of extending the familiar human form of consciousness into realms far distant from ordinary life. In some cultures the business of "consciousness expansion" is an honored profession. The persistent ingenuity that some humans have shown in inventing techniques for altering awareness suggests that the urge for inner exploration is as fundamental a drive in human beings as the urge to explore new physical frontiers.

Activities and substances used in techniques for modifying our familiar form of inner life include meditation, mantra, chants, dancing, incense, music, and fasting; psychoactive roots, stems, fruits, and seeds; LSD, DMT, and MDMA; yogic postures, martial arts, and spirit possession; no sex (celibacy), slow sex (tantra), stored sex (coitus reservatus), and horde sex (orgy); vision quests, carbon dioxide inhalation, sensory deprivation, and sensory overload; whirling, marathon running, hypnotism, and repetitive prayer; marijuana, mass rallies, solitude, and mutual gaze; carnival, shamanic trance, massage, and childbirth; self-inflicted pain, fermented grain, poetry, and encounter groups; strobe lights, nitrous oxide, drumming, and ceremonial magic; gestalt therapy, religious ecstasy, hot baths, and cold showers. Whatever other kinds of minds may need, the craving for self-transcendence seems to be a prominent feature of the human style of awareness. Commenting on the human need to escape the ordinary, Aldous Huxley claimed that the natural rhythm of human life is routine punctuated by orgy.

After the experimental difficulty of measuring the presence of inner experience in other beings, the next major barrier to creating a true science of consciousness is the lack of an adequate theory of how beings manage to gain consciousness and lose it. What is actually happening when I fall asleep?

Theories of Inner Experience

It is not that we possess bad, partial, or flawed theories of the inner life. We have no such theories at all, even bad ones. Instead we possess only vague fantasies, philosophical hunches, and speculative, untestable guesses. Make no mistake: we are in the kindergarten, sandbox stage of consciousness research. We have a long way to go before we can call what we know about the inner life a "science." Kurt Vonnegut's fictional Tralfalmadoreans (*Sirens of Titan*) accurately assessed the current state of human awareness research: "They could not name even one of the fifty-one portals of the soul," the aliens reported.

Scientists can say that the phlogiston theory of combustion is wrong because they have a modern theory of heat by which it can be judged and found wanting. However, present-day science is not in a position to judge claims of spirit communication, out-of-body experiences, reincarnation, telepathy, and other unusual styles of awareness systematically since it does not possess a theory of ordinary consciousness, let alone its variations. At this stage of our ignorance, scientists like everyone else must appraise these unusual mental experiences from their own cluster of amateur notions.

Long before the hundred-odd chemical elements were isolated and named, the ancient Greeks devised a rough picture of the world's fundamental constitution appropriate to their limited knowledge. They considered the world to be made up of four elements: Earth, Air, Fire, and Water, while the heavens contained a fifth element, Ether (or "quintessence") not present on earth. The ancient philosophers attempted, like scientists today, to reduce the world's bewildering variety to simple interactions between a few basic components, reducing nature to a kind of language written in a comprehensible elemental alphabet. It is ironic that our present picture of matter recognizes none of the ancient categories as truly elemental. Today's chemist regards more than one hundred substances as elemental. Particularly important "letters" in the chemist's alphabet are hydrogen, oxygen, nitrogen, carbon, and phosphorus, the five elements most essential for life.

The physicist digs deeper, breaking the chemist's elements into more fundamental parts. In what is called the *Standard Model*, present-day physicists are able to describe all known physical phenomena (conscious phenomena are explicitly excluded from the scope of physics) correctly with only three types of fundamental particles. Leptons and quarks form the "bricks" of the material world, and particle/waves called gluons form the "mortar" that sticks quarks and leptons together.

The first step toward a true theory of consciousness is to construct a rough map of the intellectual territory that we intend to explore. At this stage our maps of mind will be at least as crude as the Greek five-element picture of the world, but one must start somewhere.

At first glance the world seems to consist of two kinds of phenomena: mental experiences and physical objects. The sixteenth-century French philosopher René Descartes called these two categories *res cogitens* (thinking stuff) and *res extensa* (extended stuff—stuff that occupies space). The Greeks dubbed these two basic essences "psyche" (ψυχη) and "physis" (φυσις), from which we derive psychology, the science of mind, and physics, the science of matter.

Philosophers call the question of how mind and matter are related the "mind/matter problem," the "ψ/φ problem" (pronounced "psi/phi"), or alternatively the "mind/body problem," since animal bodies like our own are the only presently known vehicles for the occurrence of inner life.

Given such a two-component world, it is easy to work out (and make up names for) all the logically possible relations that might exist between mind and body. Either the mind/matter distinction is fundamental—neither component can be reduced to the other, a philosophical option called *dualism*—or the mind/matter distinction is only apparent—the world really consists of only one fundamental substance, a situation called *monism*.

Dualistic Models of Consciousness

A dualist maintains that mind and matter are essentially different kinds of essences each with its own laws and manner of existence. Some dualists speak of a "soul" that inhabits and enlivens the body, a sentient essence that may even survive the body's death and dissolution. In this materialistic age, dualists are often accused of smuggling outmoded religious beliefs back into science, of introducing superfluous spiritual forces into biology, and of venerating an invisible "ghost in the machine." However, our utter ignorance concerning the real origins of human consciousness marks such criticism more a matter of taste than of logical thinking. At this stage of mind science, dualism is not irrational, merely somewhat unfashionable.

There are essentially three kinds of dualism, depending on which of the two partners in the mind/matter marriage is seen to have the upper hand.

In *epiphenomenalism*, matter is the real substance of the world and mind a mere byproduct completely subject to matter's motion. Matter and mind interact but the interaction is a strict one-way street with mind as slave, matter as master. In this view, mind is like the light that goes on when you throw the (matter) switch. The switch controls the light; the light never controls the switch. There is probably no better motto for epiphenomenalism than that of the nineteenth-century Dr. Vogt: "The brain secretes thoughts like the liver secretes bile."

On the other hand, if we imagine that behind every material motion lies an invisible spiritual cause, then matter is wholly subordinate to mind—the *animism* hypothesis. To a wholehearted animist every material thing is alive and possesses a soul, which rules its external behavior. A philosophically minded animist might justify his belief by appealing to the behaviorist discovery that no conceivable human behavior can reveal with certainty the presence of an inner life in another person: from the outside human beings appear to be "mere things." But we know (by private revelation) that these apparent things actually possess lively insides that control their behavior to some extent. Therefore, it seems plausible that many other apparent things—trees, rocks, stars, and spiral nebulas—may possess similar insides. The animist is an open-minded soul who is willing to grant the gift of inner life not only to humans and so-called higher animals but to every arrangement of matter in the known universe.

From our experience as embodied beings, we know that body states can powerfully influence our states of mind. Likewise intentions that seem to originate in the mind can control the body's movements. It seems, at least in the case of human-style awareness, that neither matter nor mind dominates the mysterious partnership that gives rise to our external actions and internal experiences. This evenhanded form of dualism in which mind and matter mutually influence one another is called *interactionalism*. Critics of dualism have questioned how an entity that has no spatial location can interact at all with a body that occupies space: how does the mind find its body? Although this question might be answered in many ways, some of which we will consider later, it points out the great disparity that exists between our knowledge of matter and our knowledge of mind. Although we know of matter only secondhand through the mediation of our senses, we have managed to develop an elaborate mathematical understanding of this indirectly known essence, an understanding that extends from the tiniest elemental quark to the entire universe.

On the other hand, although our experience of consciousness is direct and unmediated, we possess no equivalent physics of mind, no elaborate conceptual structure that mirrors the rich mix of inner experiences enjoyed by human beings. Our fledgling mind science—psychology—has produced only fragmented accounts of particular aspects of human personality and seems far from achieving a comprehensive model of consciousness that is explicit enough to connect conceptually with our very detailed model of matter. When matter interacts with mind, just what kind of entity is it encountering? A good dualistic model of the mind would not only describe the nature of mind-in-itself, essentially a map of the soul, but also take up in great detail those attributes of matter, those qualities of soul, that permit these two fundamentally different aspects of the world mutually to affect one another.

Monistic Models of Consciousness

Monism, like dualism, is of three main types, depending on which of the two primary essences is elevated to the status of grand monarch.

In *materialism,* matter is all that there is. Democritus, the early Greek atomist, said it best: "By convention sour, by convention sweet, by convention colored. In reality, nothing but atoms and the Void." Epiphenomenalism, although it makes mind subordinate to matter, at least grants mind a separate existence apart from matter. But, for the materialist, mind has no special status: it is just one of matter's possible attributes, on a par with momentum, energy, and center of gravity. In his splendid book *The Psychobiology of Mind,* William Uttal, a modern materialist, declares: "Mind is to the nervous system as rotation is to the wheel."

Although materialists agree that mind (defined as "inner experience") is nothing more than a particular motion of matter, they differ concerning how complex matter's movement must be actually to produce a noticeable sensation, to generate what might be called a "quantum of sentience," a mental quantity analogous to physicist Max Planck's famous quantum of action—the least amount of mechanistic interplay that physics permits.

Reductive materialists believe that virtually any mechanical motion results in some kind of inner experience. Just as all particles possess momentum and energy, so all particles possess a bit of inner life. For such broad-minded materialists even atoms are conscious, although the inner experiences of such simple mechanical systems would be minuscule compared to the inner lives of human beings. Reductive materialists resemble animists in their willingness to believe that everything is alive. However, the materialist holds that inner life does not exist as a separate immaterial soul, but is a purely mechanical property that at present we lack the tools to measure. Ardent materialist Thomas Henry Huxley, Aldous Huxley's grandfather, predicted that just as heat was discovered to be nothing but a form of mechanical motion (the mechanical equivalent of heat is 4.185 joules per calorie) so likewise mind will be found to be a form of mechanical motion whose mechanical equivalent (so many joules per cogiton?) will be measurable some day by some future psychophysicist.

Emergent materialists also believe that consciousness is a wholly mechanical property of matter, but that only very complex systems possess it. To an emergentist, consciousness is less like the attributes momentum and energy—properties common to all mechanical systems—and more like the attribute "capable of producing speech," possessed only by certain special mechanical systems (speech synthesizers, tape recorders, radios, parrots, and so forth) as well as by human beings. Like the ability to produce speech, the capacity to enjoy inner experience arose through the action of biological evolution and, on this planet at least, is unique to human beings and their close relatives on the evolutionary ladder. Since consciousness is a strictly mechanical, although very complex form of motion, there is no barrier, in principle, to building machines whose quality and quantity of inner experience equal or exceed our own. Because of its simplicity, concreteness, falsifiability, and general concordance with the present fashion of scientific thinking, emergent materialism has itself emerged as the dominant mind/matter philosophy of the scientific community. When asked whether he believed that machines could think, legendary cybernetic pioneer Claude Shannon replied: "You bet. We're machines, and we think, don't we?"

Materialism is popular among scientists, especially experimentalists, whose daily lives are spent exploring the rich details of matter's lush variety, but many philosophers, whose business it is to work with less tangible stuff, have argued that not matter but mind is the fundamental substance of the world—a type of monism called *idealism*. These mind monists argue that our most direct and unmediated experience of the world is entirely mental in character. In contrast, the existence of a material world is inferred in a convincing but wholly indirect manner from evidence presented to our consciousness. The existence of inner experiences is undeniable, but to an idealist the existence of an external world is not so certain. The dream state is an often cited example of an internal experience that convincingly simulates an external reality that simply does not exist outside the mind.

The most famous idealist was probably George Berkeley, an Irish bishop for whom the California university town was named. William Butler Yeats recalled the mind-centered philosophy of his idealistic countryman in these lines:

> And God-appointed Berkeley that proved all things a dream
> That this pragmatical, preposterous pig of a world,
> its farrow that so solid seem,
> Must vanish on the instant if the mind but change its theme.

Idealism seems a foolish intellectual pastime in a materialistic age such as our own because it dismisses as illusory the material sphere in which we have made our greatest cultural progress, without offering any practical program for advancing our knowledge of the world. On the other hand, idealism suggests the possibility of developing a wholly mental science based on the manipulation and observation of states of consciousness rather than states of matter. (Some Eastern thinkers claim that such a mental science already exists.) If the material world, as the idealist claims, is like a movie being projected from a mental "projection booth," then scientific mastery of the projection mechanism could render our vaunted physical sciences superficial and irrelevant. In an essentially mental universe, the entire physical world would be reduced to the status of a movie: *Matter as Maya: The Only Game in Town.*

Neutral monism attempts to strike a balance between the extreme claims of both materialism and idealism. The neutral monist posits the existence of a single substance possessing both mental and physical attributes. An example of such a double-duty entity in science is the electromagnetic field, first described by Scottish physicist James Clerk Maxwell in the latter half of the nineteenth century. Before Maxwell, electric and magnetic forces were considered separate entities each with its own laws. Looking deeper, Maxwell showed how electricity and magnetism could be understood as interrelated manifestations of a single electromagnetic field. Maxwell's discovery was the first instance of the unification of two separate physical forces into a single description, a trend that continues today as physicists search for a grand unified theory (GUT) that will unite all of nature's fundamental forces under a single banner.

In the other materialistic models of mind, a system's inner life is entirely determined by the motion of matter. In a certain sense, it *is* that motion, just as sound *is* a vibration of air molecules. In the neutral monist account of reality, matter and mind are interdependent aspects of some more comprehensive kind of substance that includes them both as special cases. The neutral monist model predicts that a purely physical account of the world must be factually wrong when it attempts to deal with systems that possess a substantial conscious component, just as a purely electrical account of charged-particle motion will fail whenever magnetism enters the picture. Neutral monists look to a truly unified field (TUF physics?) in which the powers of mind are amicably united with the powers of matter in a single comprehensive description of all natural phenomena both inside and outside.

Religious Models of Inner Life

These philosophical guesses concerning possible solutions to the mind/body problem are more than academic exercises. In one form or another these ideas determine how people all over the world regard their lives, the people around them, and their ultimate destinies. More than dusty philosophical hypotheses, these conceptual models of mind form the core assumptions of the world's great religions. Under the guise of religion, each of the four major models of mind—dualism, idealism, neutral monism, and materialism—has attracted large numbers of believers whose lives are guided (largely unconsciously) by these philosophical assumptions. Although some of these positions may seem quaint or preposterous, each of them can claim millions, and in some cases billions, of followers.

Judaism, Christianity, and Islam emphasize the importance of the individual human soul, which they consider to be separate from the body. In these dualistic religions the body is generally seen as inferior to the soul if not downright evil. Many dualists believe the soul to be immortal, surviving the body's eventual death and decay. Although differing greatly in details, these dualistic creeds generally agree that the soul's goal is to escape matter's menial constraints and seek union with God, who is in some unimaginable sense a Person like us.

Hinduism and Buddhism view the material world as a kind of illusion. The real reality is mental, called *Braman* by the Hindus, "consciousness" or "Big Mind" by Buddhists. Although they differ concerning the strategies one should follow to become aware of the illusion, and what one should do (or not do) once one has pierced the veil of Maya, these religions basically agree with the idealistic Bishop Berkeley that the world is more like a sleeper's dream than a solid atomic drama.

Like the philosophy of neutral monism, Taoism is based on the belief that the world inside and outside consists of one substance called the *Tao*, the "Way" or the "uncarved block" from which all phenomena both mental and physical draw their existence. Mind, matter, the self, and external objects are not separately existing entities but are incomplete aspects of the single Great Way viewed from a limited human perspective. The Taoist's task is to discover the presence of this Way in herself and to learn to live in harmony with the Way's meanderings.

Materialism as a hypothesis forms an important part of the scientific enterprise; materialism as a "religion"—an unreasonable faith in reason itself—is another matter. Atheistic materialism is an active unbelief in God, soul, afterlife, or any other spiritual concept that cannot be completely anchored in a model of the world made solely of matter and ruled by the impersonal laws of physics. The materialist's goal is to pursue happiness stoically in whatever forms he finds agreeable until death definitively ends his quest.

Because these philosophical positions are woven so deeply into religious thinking, new discoveries in the science of mind are likely to challenge many deeply held religious beliefs. If these mind/matter models someday become open to experimental investigation, then beliefs that were once a matter of church doctrine or personal faith could be established as a matter of public knowledge. Experimental facts concerning the existence and nature of the afterlife would be particularly revolutionary.

Criteria for Consciousness Theories

A good theory of consciousness must be more than a plausible story or philosophical language game. The enormous success of the physical sciences provides us with high standards by which to judge candidate theories of the world. In particular the partnership between mathematical theory and sophisticated experimentation has given physicists a solid basis for their claim that they really "understand" the material world at every scale from the whole universe down to the smallest quark. The fact that physics theories are expressed in mathematical language does not mean that a theory of mind must also be expressed mathematically. Bertrand Russell once said that our physical theories are mathematical not because we know so much but because we know so little: it is only the world's mathematical properties that we have been able to discover. Isaac Newton, who more than any other man was responsible for developing the idea that the material world is governed by rigid mathematical laws rather than the whims of the gods, had to invent a new field of mathematics (calculus) in order to calculate the motions of the moon and planets. Perhaps some future Newton of mind science will also need to invent entirely new theoretical techniques appropriate for describing the essential features of the inner life of conscious beings.

How will we recognize a good theory of consciousness when we see it? I propose that we score fledgling models of mind according to how clearly, explicitly, and correctly they deal with twelve important questions.

1: Mind Links. How can I objectively determine the presence and quality of mind in material configurations other than my own brain? A good mind model should tell us how to build real mind links or show us what part of the optical spectrum to scan for the awareness-specific "purple glow." Or give good reasons, rather than fashionable guesses, why hopes for such objective tests are just wishful thinking. The importance of establishing ways of directly contacting other minds cannot be underestimated. Without this ability, the experimental side of mind science will be severely restricted. With this ability, the science of consciousness, based not on analogies and plausible guesses, but on a growing body of experimental facts about the inner lives of sentient beings other than us, will truly begin.

2: Mind Maps. What kinds of minds besides our own inhabit the physical world? The success of matter science has ousted us from the cozy medieval geocentric world into an almost inconceivably vast universe filled with innumerable stars, galaxies, and strange cosmic objects. Material science has numbered the chemical elements, broken them down into parts, and further analyzed these parts into truly elemental particles. Science has completely surveyed the physical universe and finds it filled with an immense variety of material forms. But what of the mental realm? Is nature's inner life as rich and various as her outer behavior? Is my heart a conscious being? My hand? Is the earth not only a self-organizing mechanism as James Lovelock's "Gaia hypothesis" supposes but also a conscious being with feelings, perceptions, and a certain freedom of action? Is the earth, in short, a person like me? Will advanced mind links someday allow us to communicate directly with the "soul of the earth"? What can we say about the inner life of atoms? A good theory of consciousness, either by supplying us with an experimental method for answering such questions or by providing absolute theoretical specifications for the presence of awareness in material systems, should allow us to construct a "geography of the mind," a mind map as rich in detail as the *New York Times Atlas of the World*, illustrating the major centers of awareness in our little corner of the material world. Until we have a better notion of the true extent of the world's inner life, we are like geologists holding one stone or biologists looking at a single living specimen.

3: Artificial Awareness. How can we construct machines that possess insides like ours? Because we have learned how the kidneys work, we can make artificial kidneys that perform the same vital function. Once we know how the brain produces (or hosts) consciousness, we should be able to manufacture beings that enjoy inner experience, or augment our own lives with synthetic forms of awareness. If, as the monists claim, matter and mind are one, then we can construct an artificial mind out of ordinary matter. Dualists, on the other hand, believe that mind comes from "outside" to inhabit matter. In that case, the best we can expect a theory of consciousness to do would be to tell us how to build a maximally attractive dwelling for eventual habitation by external sentient entities. If you wanted to build a conscious robot—like Hal 9000, for instance—what sort of parts would you order? And who would supply them: an electronics shop, a biology tank, a physics lab, or some new specialty shop stocking wares at present inconceivable?

4: Quantity of Mind. What feature of matter determines the quantity of a being's conscious awareness? It is a common experience to find it difficult to pay attention to more than one thing at a time, and not every detail of that thing can be simultaneously held in mind. There seems to be an upper limit to the amount of attention we can muster. And compared to the vast number of sensations, thoughts, and feelings craving attention, this amount of focused awareness (we will estimate it later) seems to be quite small. In states of sleep, coma, and general anesthesia this attention rate drops to zero, and we lapse into a state correctly called "unconsciousness"—no inner experience whatsoever. A good theory of consciousness should explain how consciousness is extinguished: how sleep, coma, and deep anesthesia are produced in the human brain (and in other sentient systems), how awareness is reestablished after such unconscious interludes, and what physical or spiritual parameters determine the magnitude of our "conscious data rate."

Another quantitative feature of human awareness (and probably other styles of awareness as well) is what psychologist William James called the "specious present." Our awareness occurs not as a succession of single instants but as a flowing together of extended periods ranging from a fraction of a second to several seconds in duration. What material mechanisms determine the subjective length of the "present moment"? Once these quantitative questions are answered, we can then apply artificial awareness techniques to literally "expand human consciousness" in both the temporal and the data-rate dimensions.

5: Mind Quality. What determines the *quality* of conscious awareness? How in the world is the smell of cinnamon produced? The color green? The sense of vertigo, the taste of peppermint, and the sound of music? How can mere matter feel pain and pleasure, fear and anticipation? Do there exist new colors, tastes, completely novel senses, emotions, and modes of being that we may be potentially capable of experiencing, but that our present biological makeup does not support? What are the dimensions of "experience space"—the realm of all possible experiences open to a conscious being?

Some people (Douglas Hofstadter, author of *Gödel, Escher, Bach*, for one) believe that the essence of consciousness is the ability to self-reflect. Others (Immanuel Kant comes to mind) put moral and ethical abilities in first place. It seems to me, however, that these features are luxuries possessed by our familiar human form of awareness. Such features would not necessarily be present in primitive, memoryless forms of inner life. A good consciousness theory should be able not only to account for the qualities of inner life available to beings more primitive than us but to allow us to extrapolate to higher forms of awareness not yet experienced by human beings.

6: Attention Mechanisms. How does matter "pay attention"? Besides the crude distinction between outer behavior and inner life, it seems plausible to make a second distinction between the active and passive qualities of inner life. Although it surely contains some active elements such as the focusing of attention, the process of perception seems to serve a primarily passive, receptive function. Perception is something that happens to us; the perceptual field floods us with surprise, with sensations mostly not of our own making. On the other hand, the experience of voluntary action is primarily active, controlled by something in the mind, although it surely contains passive elements such as stereotyped or unconscious behavior patterns. A good theory of awareness would explain the phenomenon of active attention: its "motion" from one experience to the next against a background of passive, unattended-to, automatic activity.

7: Sense of Self. One of the most striking features of our human style of consciousness is its unity. Even in pathological cases of multiple personalities, only one personality at a time takes control. The mind's felt unity is all the more surprising when we look at the brain, which is not a one-operation-at-a-time machine like a computer but a massively parallel processor in which millions of interconnected events are going on at once. Perhaps there are "crowd minds" somewhere in the world that experience several flows of awareness simultaneously, but humans seem to have been put together with decidedly one-track minds. Related to this human singleness of being is the familiar "sense of self," the feeling that one enduring being is enjoying these experiences, a being who changes somewhat from moment to moment, and from day to day, but remains in essence the same person. Beliefs concerning the nature of the self range widely, from the Christian idea that self is an immortal being to the Buddhist claim that self is nonexistent, a mere illusion. A competent theory of mind should explain the human mind's singleness of being in the midst of the brain's multiplicity of functions and resolve as well the vexing question of selfhood: is self an illusion or not?

8: Personality. How can we best describe an embodied being's inner traits and range of possibilities? What is the essence of personhood? Systems as diverse as astrology, psychology, and occultism have produced a variety of models of personality: Are you a Gemini, a Myers-Briggs ENFP, or an Enneagram Apex 4? A good mind model should generate a theory of personality—both human and robotic—based not on external behavior but on the structure of the material (or spiritual) processes that support the inner experiences that form personality and character. How many ways can inner experience be organized into relatively autonomous units? Just as question 5 asks for a catalog of possible experiences, so we should also ask for a catalog of possible ways these experiences can be structured into wholes. For instance, can dual or triadic beings exist whose inner lives are organized around more than one central core? What are the conditions necessary for forming a "person" out of isolated inner experiences? What is the material basis for personhood?

9: Free Will. Associated with the sense of self and personality is the notion of "free will." Although much of my behavior is unconscious, automatic, or reflexive, I have the feeling that some of my activity is not forced upon me but results from free choices made by my "self." Concordant with this belief that the self is in charge of its behavior, most legal systems hold a person responsible for his acts except in situations where he can prove that he acted under irresistible compulsion. The philosopher Spinoza, on the other hand, dismissed free will as an illusion resulting from our ignorance of the true causes of our actions. A good theory of consciousness should be able to resolve the free will question by revealing the ultimate causes of our willed acts: do these causes reside solely in matter or do they originate in an immaterial soul? What does it really mean for an action to have psychological rather than material causes? Do willful acts violate the laws of physics?

10: Mr. Death. No altered state of consciousness is more drastic and inevitable than that which accompanies the body's death and dissolution. Speculations concerning the fate of the inner life at the moment of death include absolute extinction, entry into an afterlife, merging with a larger Mind, or reincarnation into another body. A theory of consciousness would be manifestly incomplete if it could not resolve on scientific rather than religious or philosophical grounds the important question of what actually happens to the mind when the body ceases to exist.

11: Mind Reach. Many parapsychologists claim that the mind can both sense and influence the material world and other minds outside the usual sensory and muscular channels. These alleged extrasensory powers include telepathy, psychokinesis, clairvoyance, psychometry, distant healing, and distant influence of other minds. A good model of inner life would provide both a framework for understanding such phenomena and an explanation for why humans are able to exercise such useful powers only very infrequently. From a scientific point of view, a theory's most desirable feature is its falsifiability: a good theory makes bold predictions that could turn out to be wrong. On the other hand, a bad theory makes vague predictions, forecasts that cannot be put to the test, or comes up with an explanation no matter how the experiments come out. Concerning the existence of extrasensory powers, the emergent materialism mind model takes an admirably bold and falsifiable stand: it predicts that all such powers are utterly nonexistent. Rival models of mind that have room for such powers have not been developed to the point where they can set testable limits to the mind's alleged extracorporeal reach. Although ordinary awareness is abundant, undeniable, and readily accessible, crucial tests of rival mind models may be better carried out in the rarefied realm of these extrasensory extensions of ordinary experience.

12: Evolution. Science accounts for the abundant variety of lifeforms on this planet by a process of natural selection operating on population diversity created by genetic variability: only the fittest survive. This evolutionary perspective teaches us that only biological traits that have survival value will endure the rough-and-tumble genetic lottery. Consequently we must ask, What is the survival value of consciousness? How did the possession of an inner life (compared to existence as a skilled but wholly unconscious automaton) aid our ancestors in their struggle for existence? At what point on the evolutionary scale does consciousness emerge? Or is the inner life a feature of the world that lies outside the evolutionary story, for whose origin we must invoke higher principles than natural selection?

13: Surprise. The most important feature of a good theory of consciousness might not be how well it explains presently known or half-suspected properties of human awareness but its disclosure of previously unknown, or even undreamt of, phenomena that have remained invisible for thousands of years, obscured by our own ignorance and lack of imagination. As the science of physics matured, it disclosed hundreds of new particles, physical effects, and invisible realms of being, and it continues to do so. We should demand no less of a scientific theory of inner life. We should expect mind science to open our eyes and hearts to unexpected possibilities of being, expect it to surprise us in magnificent ways that we could never have foreseen.

Compared to our knowledge of the physical world, our understanding of consciousness is minuscule. The major drawback to a science of the inner life is the stubborn fact that consciousness is invisible: we cannot see, hear, feel, or taste it. Since science is based on knowledge gained through the senses, consciousness is publicly accessible only indirectly. The behaviorists could even boldly deny the existence of consciousness, and science, to its shame, could not prove the behaviorists wrong. Though there is no known behavior that is consciousness-specific, we do know (by private revelation) that consciousness certainly exists. Novels, plays, poetry, opera, and other works of art explore in rich detail what it's like to be a (human-style) conscious being. Even science is not entirely powerless to gain accurate information about the inner life of humans (and some animals). Using indirect methods, ingenious psychologists have been able to map certain major features of human awareness, uncovering a body of objective facts about subjective experience.

consciousness from inside: prominent landmarks of inner space

The soul may be a mere pretense
The mind makes very little sense
So let us value the appeal
Of that which we can taste and feel.
—PIET HEIN

Please let this feeling last.
—POPULAR SONG

In the early sixties a group of psychologists calling themselves the "Third Force," to distinguish their movement from psychoanalysis and behaviorism, rediscovered the importance of the body. A flood of body-centered therapies emerged to soothe troubled minds, including somatic manipulations associated with the names Wilhelm Reich, Moshe Feldenkrais, Milton Trager, and Ida Rolf. To redress conventional psychology's alleged overemphasis on intellect, the formerly neglected body was treated to courses of bioenergetics, deep tissue work, dance therapy, martial arts, and various styles of massage. A popular practice at that time was "sensory awareness," which involves reacquainting yourself with your body by systematically focusing total attention on one body part at a time. Since most of us usually take our body's parts for granted unless they are in pain, simply giving these neglected parts full attention can be particularly invigorating and often leads to personal insights about the way our selves are habitually embodied in the world.

I remember lying on the lawn at Esalen Institute one summer afternoon, with Bernie Gunther, one of the deans of sensory awareness, exorting us to "Come to your senses." During this exercise I became particularly conscious of the fact that although I could bring to mind each of my fingers, I could not do the same with my toes. To my conscious mind, my feet seemed to be undifferentiated lumps, like sacks of beans. After this experience on the Esalen lawn, I resolved to go barefoot more often, to free my feet from their hard leather prisons.

I was talking about sensory awareness with my companion Claire, a humanoid entertainment robot freed by the Electronic Emancipation Act of 2050. As you probably know, all robots take periodic Turing tests in which they attempt to simulate a wide variety of apparently inner-directed behavior. A graduate of the cybernetic equivalent of Bennington, Claire had no difficulty in passing her T tests with honors. I asked her whether robots ever indulged themselves in sensory awareness. Could she, for instance, concentrate all her awareness into her right middle toe? Claire furrowed her brow, quivered her lower lip. "Don't you understand," she sobbed. "I can do anything a human can do, and lots of things that humans can't [here Claire's electric eyes briefly twinkled]. But my inner life is nil, a complete zero. Any awareness I may seem to have is just your own projection. I'm nothing [sob!] but a clever fake."

I put my arms around Claire to comfort her. "You're such a lovely fake, Claire. But why are you crying?"

"Because that's the way I'm programmed, you idiot!"

When she had calmed down, Claire explained that the concept of robotic psychology was born out of the detailed mappings of human inner space begun in the twentieth century, plus the development of compact "as-if" circuitry capable of computing the behavioral consequence of any precisely specified inner state. Simulating emotion was particularly easy, Claire continued, because of Plutchik's discovery that the emotions of biologically based beings stem from eight basic internal forces, or impulses. Engineers and scientists had always had problems in dealing with emotions, but forces were something they felt that they understood. Once passion-sensitive psychologists succeeded in mapping these inner forces onto the appropriate "emotional space," the mechanics of desire were quietly worked out and quickly realized in electric circuitry.

After her speech I was moved to ask, "Claire, do you desire me?"

"No, not really, Nick. But I think I can work up a pretty good simulation."

Unlike honest but heartless Claire, we all know what it feels like to be a conscious being. Consciousness, whatever else it might be, is a certain kind of inner experience. This lively, presently private, interior drama has certain explicit features that I outline in this chapter, but the rock bottom fact about awareness is its very feel, the tang of being. So familiar is this state of existence that it is hard to confront, like fish trying to feel water. Mind can be examined, but self-examination never catches human awareness in its natural state, seeing only mind-under-scrutiny complete with a scrutinizer. No matter how lightly the mind tries to touch itself, the unexamined inner life is not open for introspection. This awkward, self-referential impasse, home base to the meditator and introspective philosopher, may impose ultimate limits on what can be learned about awareness through self-examination, but the rough characterization of human awareness presented here will not press these limits. Most of our experience most of the time does not involve self-reference at all. There are so many other things in the mind clamoring for our attention that the mind itself as a topic of contemplation usually gets low priority.

Since all the contents of consciousness—every concept, experience, perception, emotion, or memory that we can bring to mind—are already saturated with the "tang of being," it may be difficult to find a neutral experience in the mind itself, in terms of which the mind might be understood. Logically the most obvious way to explain consciousness would be to show how conscious experience can be built of material that is itself unconscious. Although our lives are punctuated by intervals of unconsciousness, we have no private insight into the unconscious state because by definition such a state cannot be experienced. We close our eyes to fall asleep, and then (barring dreams) morning comes immediately. There is a sense in which we are always awake, and always have been. Nobody remembers, or can remember, not being conscious.

The Experience of Losing Consciousness

Consciousness is our most valuable possession—and more than a possession, the possessor as well, the self, the central sun, the precious ego for whose needs we daily toil—but each night in a familiar ritual we willingly give the ego away and casually surrender ourselves to a period of self-extinction. If sleep were not such an everyday process, we would regard it as an awesome mystery. Suppose we were beings that slept only once in our lives, sometime around the age of 30 years. Society would then be divided into the Old—who have survived the Experience without Experience—and the Young, who have not. Imagine trying to explain to your son what it is going to be like for him not to have a self for a period of 8 hours.

Although consciousness is regarded as a "higher function" of the brain, compared to more mundane operations such as regulation of the heart and lungs, this exalted function is extremely vulnerable to external conditions—usually it is the first function that the body abandons when the organism is stressed. Low blood sugar level, loss of oxygen, excessive bleeding, concussion, inhalation of certain anesthetic gases all reduce the brain (as far as we can tell) to the status of an unconscious machine. This so-called higher function is so tentatively linked to its more robust body that even standing up too fast can cause some of us to faint, a phenomenon called "blood pooling" in which the brain's blood supply temporarily decreases, as the heart adjusts to the increased hydraulic demands of the upright posture. Blood pooling provided me my earliest experiences of an altered state of consciousness. As a child (and sometimes even today) whenever I got up fast and felt this strange state coming on, I would hold on to something solid and try to savor the experience, which in extreme stages involved temporary blindness and brief, terrifying episodes of ego loss. I never sought out this state—it is not entirely enjoyable—but welcomed it whenever it presented itself. Although blood pooling is an experience easily accessible to people with the proper physiology, I have yet to meet another connoisseur of this particular path to altered states of consciousness.

Certainly one of the main research areas for a science of consciousness continues to be the elucidation of the mechanism of sleep, of general anesthesia, and of other physical processes that lead to the abolition of (our type of) awareness in the brain. One of the major clues as to what causes human consciousness to be present in the brain is the nature of those operations that reliably produce the absence of consciousness. These crucial events whereby a being with inner life turns into a soulless robot offer the awareness researcher a class of inner phenomena with the greatest possible existential contrast: the difference between full consciousness and no consciousness whatsoever. How do these pivotal transitions occur? And how essential to the world's affairs can a mind/body connection be that is interrupted so easily: by a tap on the head or even a hop out of a hammock?

The transition to the unaware state is not always abrupt. People can learn to prolong the interval between waking and sleeping—the so-called hypnogogic state—and enter into a more loosely organized form of consciousness in which ideas, images, and emotions combine in new and unusual ways. In this twilight state, participants often find their hypnogogic imagery charged with a peculiar sense of deep significance. These "significant" recombinations of psychic content are also available under light doses of general anesthetics such as nitrous oxide (laughing gas). However, like the contents of dreams, these creative rearrangements of the contents of consciousness are usually fragile, elusive, and difficult to recall upon return to normal awareness.

Turn-of-the-century physicist Ernst Mach characterized normal awareness as "one great porridge of experience." And indeed human consciousness does feel more like a porridge— warm, soft, sticky, and associated with hungers of various kinds—than like cold, hard, isolated, emotionally neutral computer operations. Although the contents of consciousness at times can consist of certain clear perceptions and distinct ideas, these seem to emerge out of an indistinct psychic background, an active ambiguous flux at the center of our being.

Concerning this inscrutable background, one might suppose that "consciousness is always consciousness of something; there is no such thing as consciousness without content." And indeed, whenever I ask myself what I am attending to, the examined mind always seems to be "paying attention" to something. However, I am less sure that when I am not interrogating my inner states, my mind is aware of any particular "things" at all. The true state of unexamined awareness may be closer to consciousness without (explicit) content. Try to recollect what your experience was like before you began self-examination, taking into account the kinds of experiences that are easy to remember and those that are almost impossible to recall. Now try to say honestly whether your consciousness must always have an explicit content, or whether consciousness often just fades into the background, letting events flow by without grasping them.

The Unity of Conscious Experience

What is consciousness per se? It is not easy to put into words. Books on meditation point to this experience as do treatises on phenomenology. Fortunately, to appreciate this experience we do not have to depend on secondhand accounts. Each of us can enjoy for himself/herself this common ineffable state of being. Ask yourself right now: "What does it feel like to be a conscious being?" This is what consciousness feels like while it's being scrutinized, a rather rare occurrence for most people. Now ask a harder question: "What did my mind feel like before I asked these questions?" Now you are touching this book's real subject matter—the everyday unscrutinized awareness of humans and other sentient beings.

One of the most surprising features of human consciousness is the fact that although we know the brain to be a massive parallel processor with many billions of operations going on at the same time, our inner experience seems to possess a single center: whatever is going on seems to be happening to only one being.

Computer scientists call a machine that performs one operation at a time a "von Neumann machine" after John von Neumann, the Hungarian-born polymath whose ideas guided the development of the early "thinking machines." Almost all present-day computers work this way. The few parallel computers capable of performing a great many operations simultaneously are referred to as "non-von machines." Introspective evidence alone—the fact that we always experience consciousness as a one-track mind—might lead us to conclude erroneously that the brain operates as a von Neumann computer, handling only one operation at a time.

Even in pathological cases of multiple personalities, only one personality at a time seems to dominate the mind. Although one could imagine the several personalities sharing the hapless victim's consciousness in a type of psychic oligarchy, such a possibility never in fact seems to be realized. Instead the various personalities always take turns at the helm, ruling the psyche in a kind of serial monarchy. The apparent difficulty of establishing a multicentered form of consciousness in the brain may be an important clue as to the nature of awareness, or it may only be a feature peculiar to the familiar human form of consciousness. The experienced unity of human consciousness was one of the inner facts that moved Descartes to site the "seat of the soul" in the pineal gland. He noticed that this centrally located pine-nut-shaped organ is one of the brain's few unpaired structures, and hence might be a suitable control post for maintaining a unified sense of self.

Can we, by taking thought, split our awareness into two parts? We have all had the experience at a crowded party of listening to two conversations at once. However, we seem to accomplish this feat not by dividing our awareness in two but by rapidly switching our single-minded attention from one speaker to another—a process analogous to "time sharing" on large multiuser von Neumann–type computers.

Psychologists have made many attempts, some simple, some drastic, to disrupt our familiar unity of consciousness, but to no avail.

One experiment involves feeding a different spoken message into the right and the left ear. The subject must write down or repeat aloud the message he hears in his right ear. Can he "split his mind" and also attend to the message coming into his left ear? In general, he cannot, although he will usually respond if his name is called out in the unattended channel. Our ordinary one-track mind apparently cannot be divided by so simple a tool as a two-track Walkman.

What about radical surgery? The two cerebral hemispheres of the brain arise from a single stalk—the brain stem—through which communication passes to and from the rest of the body. Communication between the left and right cerebral hemispheres travels mainly through four neuronal "cables" that link up the two halves of the brain. Three of these cables are rather small, the *anterior* and *posterior commissures*, located at the front and rear of the brain's central ventricle, and the *massa intermedia*, which crosses the central ventricle to link the right and left thalamus through which messages are relayed between the cerebral hemispheres and the brain stem. In addition to these three minor connectors, the hemispheres are linked by a broad band of tough tissue called the *corpus callosum*, which consists of a colossal number of nerve fibers—about 20 million (*callosum* is Latin for "calloused").

In order to prevent certain kinds of epileptic seizures from spreading across the brain, neurosurgeons have opened up the skull and cut all four of these transhemispherical neuronal cables—the celebrated "split-brain" operation for which Canadian surgeon Wilder Penfield received a Nobel Prize. As a result of this drastic surgery, neither the patient's behavior nor his subjective experience seems to change in any major way. In particular the patient never reports experiencing a two-track mind. Radical surgery can split the brain but has not yet succeeded in dividing the mind.

Although the patient's behavior in everyday tasks seems unchanged by the split-brain operation, more subtle experiments can elicit unique split-brain phenomena. For instance, using certain kinds of goggles, two different pictures can be flashed to each hemisphere, say an apple to the left brain, a teacup to the right. The left hemisphere—which controls the speech facility in most people—can verbally identify the apple. The person says that he sees only an apple. However, when asked to pick up the visualized object with his left hand, he will lift the teacup not the apple.

To interpret this experiment it is important to remember that the left side of the brain controls the right side of the body, and vice versa, and that for most people the speech center is located in the left (or "dominant") hemisphere. These experiments seem to show that the split-brain patient's head contains two separate hand-eye control systems, each of which can recognize objects, learn tasks, and obey commands entirely independently of the other. Only one of these systems (left brain) has the power of speech; the other system is mute. The speaking system (system 1) asserts that it is conscious and claims that it is unaware of the operations of system 2.

Concerning the consciousness question, these kinds of split-brain experiments have divided psychologists into two camps. One group believes that only the dominant hemisphere is conscious—that is, enjoys an actual ongoing inner experience appropriate to the behavior of system 1. The second hemisphere is unconscious and constitutes a sort of "zombie lobe"—a clever but entirely automatic neural mechanism—no mind there at all.

The second group of psychologists believes that both systems are conscious (split brain = split mind) but only one system can speak about its experiences. If these experiments are interpreted as evidence for the presence of one mute mind in the body, they raise the possibility that more mute minds than one may inhabit our body's substance. Why not a separate mind for every organ? Does the liver act any less intelligently than the brain? Lacking a mind-link technology that can reliably reveal the presence of awareness in any given system, we cannot yet scientifically decide between these two alternative solutions to the split-brain question.

The results of the split-brain experiments are surprising and have raised more questions than they have answered. If this drastic operation would simply have divided the patient's mind into two concurrent streams of awareness, we would have created an entirely new form of consciousness—a "two-track" mind. Instead the split-brain experiments show that mind is very difficult to split. Perhaps mind is a kind of "atom" in the original Greek sense of the word: something impossible to break apart.

Constructing the Present

In its single-minded pursuit of unity, consciousness strives to integrate sensations, memories, emotions, and cognitions into one ongoing inner experience. In addition to bringing together these disparate experiential modalities, consciousness also generates a consistent sense of time—a single "present moment" that frames the various activities that form the contents of awareness at one particular time. In the words of German sensory physiologist Ernst Poppel, "We sit on the present as on a saddle thrown over time."

The present moment is not instantaneous but occupies a certain finite temporal extent whose length varies according to our psychological state. This "specious present" (William James) represents the interval of time that is simultaneously present for subjective evaluation—a kind of "attention span" bridging past and future events. One way of estimating the length of the specious present is to listen to a series of regular pulses, such as those produced by a metronome, and mentally accent every other pulse, so that the pulses are not heard as a series of identical sounds but as a rhythm of pairs of alternating upbeats and downbeats. As the interval between pulses is made greater, a point will be reached where the pulses can no longer be mentally grouped into pairs, because the two pulses fall outside the span of attention. The length of the specious present measured in this way amounts to about 3 seconds. Experiments of this sort suggest that there is a sense in which our entire life is only 3 seconds long: all else is mere reminiscence or anticipation.

How finely can we divide our little 3-second lives? The shortest perceivable time division—sensory physiologists call it the *fusion threshold*—is between 2 and 30 milliseconds (ms) depending on sensory modality. Two sounds seem to fuse into one acoustic sensation if they are separated by less than 2 to 5 milliseconds. Two successive touches merge if they occur within about 10 milliseconds of one another, while flashes of light blur together if they are separated by less than about 20 to 30 milliseconds. Experimental difficulties prevent the scientific determination of fusion thresholds for taste and smell, so you will have to judge for yourself how far two smells must be separated in time before you perceive them as separate instances of the same odor. Note that the sense of sight has the largest fusion threshold. This relatively slow response of the human visual system is a boon to designers of TV sets. If our sense of sight were as temporally sharp as our sense of hearing, the framing rate of TV cameras (now fixed at about 60 frames per second) would have to be increased by a factor of 10 to achieve flicker-free TV images.

Humans consider two events "presently perceived" if their temporal separations happen to fall in the range of times between about 3 milliseconds and 3 seconds. When we achieve access to other forms of mind, one of the most important subjective features we will discover will be the temporal dimensions of their inner lives. Assuming they are conscious beings, how long is the attention span of a redwood tree, an ant colony, a helium atom, or the Andromeda galaxy?

The human mind does not perceive the present moment so much as it "constructs" it. There is, as far as we know, no master clock or central organ of temporality in the brain. In fact, signals from different sensory modalities occurring at different times in the brain are judged by the mind to happen simultaneously.

The mind's "construction of the present" is responsible for our sense of when a sound and a flash of light or a touch seem to happen at the same time. Such subjective judgments are important for athletic performances, for the skillful operation of machinery, and for the making of accurate scientific observations. In 1795 the British Astronomer Royal Nevil Maskelyne dismissed his assistant David Kennebrooke because despite repeated warnings he persisted in recording the meridian transits of stars as much as 0.8 second later than his master. Observing a star's transit time involves a human judgment concerning the simultaneous occurrence of a visual and an auditory event: the star's image crossing a hairline in a viewfinder and the tick of an astronomical clock. The Prussian mathematician Friederich Bessel examined transit records at the Königsberg observatory and found that the transit estimations of various astronomers differed systematically from one another, by a quantity that came to be known as "the personal equation." Kennebrooke could have kept his job if his boss had known that the perceived present is a subjective construction not an objective fact. "Be here now" means something slightly different for each person on earth. The realization that everyone has his own style of constructing the visual/acoustic present set off a wealth of nineteenth-century experimentation by Exner, Wundt, and Wolfe on the various personal equations associated with people's experiences of the subjective simultaneity of sensations of touch, sound, sight, and electric shocks.

Libet's Temporal Referral Experiment

One of the most unusual features of "reality construction" in the brain was discovered recently by neurosurgeon Benjamin Libet at University of California Medical School in San Francisco. Libet found that electrical stimulation of the sensory cortex—that part of the brain's surface primarily responsible for processing tactile information from the skin—did not result in conscious sensation unless the stimulation was prolonged for at least 500 milliseconds (0.5 second), an enormously long time compared to the roughly 10- to 20-millisecond transit time required for the nerve signal to travel from the touch site to the cerebral cortex. The necessity for at least half a second of cortical stimulation before a sensation was felt held true both for direct electrical stimulation of the bare cortex and for indirect stimulation via mild shocks applied to the fingertips. In both cases if the signal as recorded at the cortex was less than 0.5 second long, the stimulation was not consciously perceived. This does not mean that a skin shock has to be at least 0.5 second long in order to be felt, but only that the secondary signals produced by skin shock signals at the surface of the brain must last at least 0.5 second before the skin shock can become a part of conscious experience.

Despite his observation that the creation of a conscious

tactile experience requires at least a half-second of sustained cortical activity—a veritable eternity compared to typical neuronal response times—Libet found that his patients experienced their finger shocks "immediately," certainly not 500 milliseconds after the shock was applied. Libet's latter result is consistent with our own experience of tactile phenomena: if we had to wait 0.5 second before experiencing what we touched, our tactile sense would be virtually useless for all but the very slowest of physical activities. Typical tactile reaction times are on the order of 0.1 second (100 milliseconds)—the time it takes to perceive a touch and push a button too. How can Libet's observation that 0.5 second of neural activity is needed to build up a conscious touch sensation be reconciled with the fact that we can feel a touch and take action five times faster than the time these perceptions are alleged to require before they can become conscious?

In a series of ingenious experiments involving variously timed electrical stimulations of both the cortex and the skin, Libet was able to resolve this sensory dilemma. What seems to be going on is this: The tactile signal reaches the cortex in about 10 milliseconds and is not consciously perceived. But this arrival time is unconsciously noted in some way. Then if the cortical activity due to the tactile stimulus is not disrupted and is allowed to proceed for the minimum time adequate to produce a conscious sensation (about 0.5 second), the touch is registered as part of the ongoing flow of conscious experience. However, the touch is not experienced 0.5 second late: it is instead "referred" to the previous time indexed by the initial pulse arrival time. It is as if the initial tactile pulse sets a "marker" in the time flow, a marker that is "redeemed" if future cortical events produce enough sustained neural activity to promote that tentative touch signal into conscious awareness.

Libet's surprising results could be interpreted as an experimental disproof of the notion of psychophysical parallelism—the idea that every mental experience corresponds directly to a particular physical process in the brain. For the subjective experience of touch occurs, according to Libet's work, not at the same time as touch-induced events in the brain but long before the neuronal events that are supposedly responsible for the sensation. For the sense of touch, mind and its associated neuronal events are seemingly out of sync by almost 500 milliseconds, not parallel at all.

On the other hand, Libet's work can be regarded as a specific example in the temporal domain of the "sensory projection mechanism" so familiar to us in the spatial domain. I do not, for instance, actually experience a star as being inside my head, where the neural impulses related to that star's appearance certainly reside, but in a space far outside my body, upon the celestial sphere apparently hanging a few miles or so above my head. The entire sensual world is experienced to be "out there" not "in here." (An exception to this projection mechanism might be the case of listening to music on stereo headphones, where occasionally the music seems to be located right in the center of my brain.) Libet's work reveals that the construction of the conscious present involves a similar subjective projection, in this case backward in time rather than outward in space. Presumably the existence of a "personal equation," idiosyncratic simultaneity judgments between different sensory modalities, results from private differences in the operation of this temporal projection mechanism for the various senses.

The Experience of Paying Attention

Once consciousness comes into existence (the porridge wakes up) and begins to construct its private present moment, a process called "attention" begins to search for "contents" with which to occupy the aroused mind. The process of attention has been compared to a searchlight illuminating a tiny part of a vast and complicated cavern: those dark portions of the mind potentially open to conscious scrutiny.

Attention seems to be important for learning, for laying

down permanent memory traces, issuing in effect a kind of PRINT command that results in certain spotlit experiences being remembered while other more dimly lit experiences fade away forever. Striking events that "catch our attention" almost against our will can be recalled years later, as well as intrinsically boring events (such as the multiplication table) on which we have voluntarily, often with great mental effort, concentrated the spotlight of conscious attention.

Not all information that enters the brain is available to conscious attention. The carotid body, for example, a peanut-sized organ close to the heart, monitors the oxygen content of the blood. But try as we may, we cannot consciously access this organ's output and savor the sweet taste of our blood's oxygenation. Nor can we learn to clench and unclench the muscles of the heart or viscera: such processes are part of the body's involuntary nervous system, locked away from the sweeping searchlight of conscious attention. The boundaries between voluntary and involuntary nervous system are somewhat fuzzy. Using biofeedback techniques people have learned to control skin temperature, brain waves, and blood pressure, but no one to my knowledge has ever learned to taste his own internal blood.

The attention process is intimately bound up with the question of selfhood (Who is it that is "paying attention"?) as well as the problem of free will (Who or what decides where the searchlight of attention will be pointed next?).

Both self and will can be explored in certain systems of meditation, none so effective, to my mind, as the simple process of sitting quietly and feeling your breath move in and out of your body. Breathing is a peculiar process in that it is a vital unconscious activity that can be easily brought under conscious control. One measure of willpower might be how long you can hold your breath, how long you can consciously resist the body's own urge for self-preservation. Breathing is, in a sense, a mind/body system situated at the very boundary of the self and not-self, and as such it can give us valuable insights, mostly nonverbal, about the peculiar experiential terms of the mind/body contract. Certainly conscious robots, if we endowed them with any measure of curiosity, will practice some form of meditation, exploring from the inside flickering self/not-self boundaries of their artificially incarnated minds.

Of all the qualities of mind, the sense of self is the most difficult to describe, let alone quantify. How will we knowingly build selfhood into our conscious robots before we possess a clear notion of what such a process means in a human being? Unless we come to a better scientific understanding of our own selfhood, it is likely that at some point in their development our robots will accidentally acquire a sense of self in a manner as mysterious as the way in which it is acquired by our children.

Three or four times each second, our eyes involuntarily jump to a new line of sight, the so-called saccadic motion superimposed upon our voluntary shifting of gaze from one point of interest to another. It has been said that the most frequent decision that our body has to make (more than 100,000 times a day) is where to look next. On what basis does the brain decide to turn its eyes in a particular direction, to speak or write a particular sentence, to perform the next voluntary action, either in its muscles, lifting the lid to smell the soup, or in its mind, recalling a childhood memory?

In a very real sense, the only power we possess is the ability to direct our attention to one particular aspect of the world rather than to another. Certainly my emotions play an important role in selecting what I attend to next, but personality, novelty, aesthetics, intuition, commitment, context, and often sheer accident also take part in this decision. In addition to these factors I have the sense of "free will"—that I could, if I wished, carry out a completely capricious action, that my choice of where to place my attention is not entirely determined by internal or external forces. I have the feeling that part of my decision concerning what to do next resides in an independent self. However, like the sense of self, what it might actually mean to have a truly free will, to be the "first cause" of certain voluntary actions, is difficult to imagine. One possible meaning of this term is that the self exists, as the dualists believe, outside the body and operates the body by means of nonmaterial forces that violate the normal laws of physics governing unconscious matter. Although the self might escape physical determinism by abiding "outside the world," it would still seem to be influenced by psychological motives. What would it mean to act without motives of any kind? Random action certainly does not seem to be a desirable kind of freedom. As the philosopher Schopenhauer put it, "I may be free to do as I please but am I free to please as I please?"

How Much Attention Can Humans Muster?

Self and will—attention's source and attention's direction—may be elusive questions, but one aspect of attention easily accessible for study is the quantity of conscious attention a person can focus on an experience. In the late 1940s Bell Laboratory scientist Claude Shannon invented the notion of *information rate*, expressed in bits per second (bps), as a measure of the data transmission capacity of any communication channel. In Shannon's terms a TV channel transmits about 10 million bits of information per second, a phone line carries about 3000 bits per second, while a tom-tom (talking drum) communicates at about 10 bits per second. What is the capacity of our conscious attention regarded as a communication channel between self and world, and how does it compare to other data rates in the human brain?

Of all the senses vision is by far the most capacious channel, transmitting from each eye on the order of 100 million bits per second of information along the optic nerve through certain midbrain relay centers into the occipital lobe located at the back of the brain. From every square inch of the skin, the largest organ of the body, tactile messages pour into and up the spinal cord, then pass through the brain stem into the thalamus, where they are relayed to the sensory cortex, a narrow ribbon of brain tissue that lies just under the headband of your stereo headphones. The quantity of tactile information streaming into the sensory cortex may be as large as 10 million bits per second. From the ears, along the acoustic nerve, about 30,000 bits per second of auditory information passes into the brain stem, where it is relayed to the primary auditory cortex located near the border of the parietal and temporal lobes. Signals informing the brain about how the external world tastes and smells carry comparatively little information compared to the vision, touch, and hearing channels.

On the output side, speech is our most capacious channel, capable of transmitting on the order of 10,000 bits per second. Virtuoso piano players and expert typists operating at top speed can produce only about 25 bits per second of Shannon-style information. The use of the whole body as a signaling channel—posture, gesture, dance, semaphore—has not been investigated, but it is likely that the whole-body data rate is low compared to that of speech. Likewise the production of olfactory and gustatory signals seems to play a quantitatively small part in human communication schemes at present.

We do not know exactly how the brain processes information, but if we assume that each neural synapse corresponds to 1 bit of information, then the cerebral cortex considered as a communication channel is capable of dealing with about 10 trillion bits per second—100 billion synapses firing at a maximum rate of 100 per second. Viewed strictly as an unconscious data processing machine, the human sensory/motor system consists of relatively modest input and output data flows linked by an enormous amount of computational power. (The human brain by this estimate is 10,000 times faster than the largest supercomputers.) Of course, almost all of this neural data processing goes on below the level of awareness. What fraction of this activity can we attend to at any one time? In other words, when we "pay attention," how much do we pay? A simple thought experiment can give us some idea of the Shannon channel capacity of ordinary awareness.

Imagine a TV display that can display sixteen different colors. The number of Shannon bits in a display corresponds

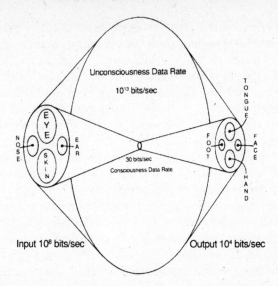

Double-funnel model of human mind/body system as information channel. Moderate input and output rates in the kilobit-megabit range are processed unconsciously at more than a trillion bits per second. In the midst of these flows, consciousness perceives and directs data flows of about 30 bits per second. Unconscious data rates outnumber the conscious rates by a factor of more than a trillion.

to the number of different possible displays expressed as powers of 2. Since 2 to the fourth power equals 16, a display with sixteen possibilities represents an information content of 4 bits. Now imagine that a colored letter can be flashed on the screen, a letter drawn from an alphabet of 256 characters (8 bits). The display now consists of 16 bits of information: 8 for character, 4 for character color, plus 4 more for background color. Can you attend fully to all aspects of this simple display: to the character itself, to its color, and to the color of its background? Probably so. Now let's imagine that the display is flashing at an increasing rate, going through all its possible changes. At what display rate does your attention fail to keep pace? Can you give full attention to the display when it is flashing three times a second—corresponding to a data rate of

48 bits per second? From experiments of this kind, it is estimated that the conscious data rate in human beings lies somewhere between 15 and 50 bits per second, much closer to tom-tom data rate than to a telephone channel. Consciousness represents much less than one part per hundred billion of the processing power of the brain. Truly our little egos are just the minuscule tip of an immense psychological pyramid.

When consciousness is taken into account, the information-rate model of the human sensory-motor system resembles a pair of double funnels. Two huge funnels throat to throat represent the coupling of sensory-motor activities to the enormous unconscious cerebral computer; two moderate-sized funnels neck to neck represent the relatively small quantitative role of conscious data processing in the human system.

How can we manage to live with such an experiential mismatch, without becoming overwhelmed or paralyzed by the body's awesome complexity? The answer lies, of course, in hierarchical organization. Although the data that enter consciousness are small in quantity, they are of very high quality. Sensory information is filtered, selected, abstracted, recoded, condensed for presentation to the limited view of conscious attention. All irrelevant details have been stripped away, the data are grouped into significant patterns, and important features are highlighted. This condensation feature makes attention look like a kind of executive in a large corporation who is ignorant of most of the billions of day-to-day details that go on in the company. He perceives his company's activities through highly condensed but relevant summaries, acts through orders whose details he leaves to his subordinates, and only occasionally ever ventures down to the shop floor. The executive handles very little information, but it is of very high quality. He gives very few commands, but they are very effective.

Can we gain a clue from this small conscious data rate concerning the location of consciousness in the brain? In the early days of brain science, when questions of the localization of consciousness were raised, an answer that was sometimes put forth (usually in jest) was that only one nerve cell—the so-called pontifical neuron—was conscious and assumed executive control over the rest of the brain. From consideration of information rates alone, one papal neuron would more than suffice to command consciously the brain's community of neurons, though more distributed, democratic theories of neural responsibility are currently in fashion.

Although quantitatively small, the contents of consciousness vary immensely in quality. These contents can be crudely divided into sensation, action, memory, emotion, and cognition. To give our robots (including poor Claire) a human-style awareness we will first have to be able to characterize the quality and range of subjective experiences available to awake, alert human beings scientifically: In other words, before giving robots minds of their own, we should get to know ourselves better, learn to construct "mind maps" that adequately represent what humans can normally bring to mind. As we discuss these mind maps, we should be aware of both how well we are equipped to interact with the external world and of the limitations of our present form of embodiment. One might reasonably expect that a future science of mind will make present maps obsolete by literally expanding our consciousness via an increased conscious data rate, by developing entirely new senses and abilities to interact with matter, and perhaps by inventing entirely new modes of being.

The biggest landmarks on anyone's map of sensation are the classic five senses: vision, hearing, touch, taste, and smell, although science writer Guy Murchie in his *Seven Mysteries of Life* counts no less than thirty-two separate senses, including a sense of balance, of appetite, of intuition, and a sense of humor. Murchie defines *sense* as any channel through which the mind relates to the body; his large sensory inventory reflects the actual richness of the mind/body connection.

Visual Space

Our primary sensory connection to the outside world is vision, a subjective appreciation of electromagnetic vibrations possessing wavelengths between 400 and 700 nanometers (1 nanometer = 1 billionth of a meter), otherwise known as "light." We do not experience vision as an unstructured blaze of light but unconsciously organize it into discrete "objects" located in a three-dimensional space at various distances from our eyes. In addition to shape, size, and texture, these objects evoke in us a certain subjective quality called "color" that seems to depend both on the object itself and on the nature of the ambient light. Color experts estimate that we can distinguish more than 100,000 different colors, which can all be mapped into a three-dimensional "color solid" or *chromasphere*, whose dimensions are conventionally labeled "hue," "saturation," and "brightness."

Hue represents the dimension of the pure colors themselves: red, yellow, green, blue, and violet. The pure colors can be arranged in a circle such that each color lies opposite its "complementary" color. A colored light added to the proper amount of its complementary color produces the neutral sensation "white." Inside the pure, or "saturated," color ring lie the unsaturated colors—pure colors mixed with a certain amount of white light. The hue and saturation variables define the two-dimensional color wheel reproduced in many art books.

The third dimension of the color solid is brightness—the visual intensity of the color experience. A black-and-white camera responds only to the brightness dimension of the color solid: this dimension is sometimes called the gray scale. One way of visualizing the color solid is to imagine that hue labels a particular color, saturation tells how much white is there, and brightness tells how much black is present in a particular color sensation.

A surprising feature of the chromasphere is the circular character of the color wheel. Since the physical variable that

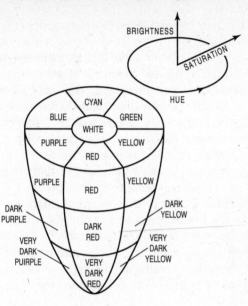

Inner-Space Color Map. Location of subjective color experiences inside a three-dimensional color solid.

corresponds to hue is the wavelength of light, and the wavelengths that human eyes respond to vary from 700 nanometers (red) to 400 nanometers (violet), one might have expected that the subjective sensation of color would likewise be spread out in a linear fashion fading away into invisibility at two limiting hues. However, unlike the physical spectrum, the visual spectrum loops back on itself, forming a color circle rather than a color line. The loop is closed via a nonspectral color purple— a particular ratio of red and violet light.

The circular character of the color wheel is explained by the fact that in normal eyes there are three different color receptors, whose sensitivities peak in red, green, and blue light, respectively. The relative stimulation of these three receptors defines a unique position within the color solid. The reason that our psychological color space appears to have more dimensions than the physical spectrum is that our eyes do not detect color as a single note, but as a three-note chord, perceiving a kind of optical harmony. If we happened to possess eyes with four color receptors, the subjective color space would no doubt be four-dimensional, the pure colors spread out over the surface of a sphere rather than a circle. Although a few poets have speculated about new color sensations outside the human range, it is impossible for us to imagine what a new color would actually look like, visually imprisoned as we are inside the normal human chromasphere.

One boon a new science of mind might be able to grant would be the literal expansion of our visual horizons, with one or more new color receptors, preferably lying outside our present visual range, whose outputs combined with those of our present receptors would allow us to perceive a vastly extended color space of more than three dimensions.

Auditory Space

The subjective color space can be mapped onto a three-dimensional solid because the eye possesses three different color receptors, whose outputs blend to produce intermediate colors. Every perceivable color can be mapped onto this space with none left out. The auditory sense is not so simple. The ear is sensitive to sound frequencies between 20 cycles and 20,000 cycles per second (almost ten octaves compared to the eye's single octave of frequencies), but the detector for these frequencies is not divided into a small number of primary receptors like the eye. Instead, the cochlea, a small snail-shaped organ inside the ear, consists of tens of thousands of tiny hairs, each sensitive to a slightly different sound frequency.

The people who design music synthesizers would like to be able to cover all of auditory space with their sound machines—to be able to duplicate all known sounds as well as to exhaustively identify all other sounds that the human ear can possibly experience. However, because the number of different auditory receptors is so large, there is no simple auditory map corresponding to the color solid upon which all possible acoustic sensations can be mapped. We have no way of simply displaying all the familiar sounds, of looking for gaps in acoustic space corresponding to sounds heretofore not experienced by humans. Consequently, there may be hundreds of novel sound sensations out there waiting to be experienced by the human ear.

In certain restricted acoustic situations, maps can be drawn of human auditory possibilities. An orchestral score is one such map. Here the number of dimensions of acoustic experience is artificially limited by the number and kinds of different musical instruments, not by the analytic capabilities of the ear.

There is a close parallel between the senses of vision and hearing because both involve sensing the frequencies of certain vibrations. Just as the sensed visual spectrum loops around, although the physical spectrum is linear, so also a corresponding acoustic loop can be produced in our subjective sense of musical pitch. By playing a three-note chord on a music synthesizer and programming the amplitudes and frequencies of this chord in a particular way, the sensation is created of a sound of constantly increasing pitch, that returns again and again to the same aural sensation—the acoustic equivalent of the visual color wheel, or of paradox artist Maurits Escher's endlessly ascending circular stairways.

Since the ear is already so receptor-rich compared to the eye, it may be difficult to expand our acoustic experience artificially, except by extending our hearing into the infra- and ultrasonic ranges.

Tactile Space

The sense of touch seems to consist of four separate senses—sensitivity to pressure, heat, cold, and pain—rather than four dimensions of a single sense. We feel pressure and heat, for instance, as two independent sensations. They do not combine like colors to produce a third new tactile sensation. Touch is our most intimate and active sense, potentially involving all the muscles of the body, in sizing up a new tactile situation. We not only passively experience changes in skin pressure but also actively engage the source of that pressure in active tactile conversation, an exploratory dermal dance resulting in complex sensations of viscous, liquid, slippery, glue; of rubbery, gritty, furry, smooth.

Our fingertips are capable of sensing a difference in height as small as 1/10,000 of an inch and can be trained to read text encoded as raised Braille dots rapidly. An exciting area of tactile research is the development of *virtual reality* simulations in which a computer generates artificial visual and tactile experiences that are channeled through a video helmet and a dataglove to produce a convincing sense of being able to sense, move, and manipulate objects in a wholly simulated environment. The dataglove senses the position and shape of the hand with magnetic and stretch-sensitive sensors and applies tactile feedback (in some designs) with an array of tiny vibrators. An exciting example of virtual-reality research is the molecular docking program at the University of North Carolina in Chapel Hill. In this simulation, the subject sees and feels a large biomolecule, which he attempts to fit by hand into its appropriate molecular receptor, an operation that taxes the computational power of the largest computers, but which is literally "child's play" for the human hand/eye combination. The attempt to produce persuasive tactile feedback for such ambitious attempts at reality simulation has spurred new interest and appreciation for the human sense of touch.

Gustatory and Olfactory Spaces

Taste and smell, our chemical senses, put us in touch with the atomic structure of the food we eat, as well as other substances that pass by and into our bodies. Taste seems, like touch, to consist of four separate subsenses, in this case sensitivity to the salt, sour, bitter, and sweet aspects of chemical substances that our tongues encounter.

The human sense of smell seems to consist of seven basic components, a sensitivity to camphoric, floral, ethereal, musky, minty, pungent, and putrid odors. John Amoore and his colleagues have shown that the first five of these basic odors correspond to the shape of the molecule that produces the olfactory sensation, and in the case of putrid and pungent, to the electric charge (+ or −) carried by the odoriferous molecule. These five basic shapes—the five aromatic solids, as it were—fit into five complementary holes in receptor sites located in the olfactory epithelium near the bridge of the nose. Our sense of smell acts as a kind of biological microscope, feeling out the shape and electric charge of invisible molecules in the air, then reporting this essentially tactile data to the mind in a peculiar olfactory code. It is interesting to speculate whether the sense of smell could ever be retrained to operate as a literal microscope, by teaching the nose to associate the seven basic smells with pictures of the appropriate molecules. To such a sophisticated smeller, a new odor might trigger off not only a new olfactory sensation, but a mental picture of the molecule responsible for the new smell.

The sense of smell in humans plays a relatively minor role compared to its importance to other animals such as dogs. Neurologist Oliver Sachs reports the unusual case of a man, who, after a blow to the head, experienced an enormous enlargement of his sense of smell. For a period of time, he literally lived in a dog's world, experiencing a dramatic tapestry of olfactory sensations wherever he went until he gradually returned to the relatively "smell-blind" human world. When we learn more about how the information collected by the sense organs is turned into sensual experience in the mind, we may perhaps all have the opportunity to "live a dog's life" if we so desire.

Other senses exist in nature whose subjective qualities we

can barely imagine. What would it be like to experience the
world via the sonar sense of a dolphin or a bat? Or sense electric fields as certain fishes do? How does a plant feel while it
is grazing on photons of light? If you could directly experience
the sizzling sensation of photosynthesis, how would you describe to someone else the taste and smell of sunlight?

When science succeeds in developing mind links that permit us to share the inner experiences of other sentient beings,
such questions will be more than academic. The mind link will
immediately rub our noses in utterly alien modes of perception. As a foretaste of what acquiring an entirely new sense
might feel like, I invite you to revive a scarcely used human
sensory ability I call "bee sight."

Bee Sight

Light from the sky is partially polarized with a direction and
intensity that vary with the sun's position. The eyes of honey-
bees are sensitive to skylight polarization, presumably to help
the bees navigate to and from their hives on cloudy days. It
is a little-known fact that human eyes can also sense skylight
polarization, but most of us never bother to exercise this ves-
tigial sense.

WARNING: once you learn to experience "bee sight," it
may be difficult for you to unlearn it. Do you really want to
contaminate your future sunset vistas with a distracting overlay of skylight polarization icons?

From a physicist's point of view, light is a transverse vibration of electric and magnetic fields that is traveling through
space at the astonishing speed of 300,000 kilometers per second. The word *transverse* means that light's fields vibrate at
right angles to the direction in which the light is moving. The
light's electric and magnetic fields also vibrate at right angles
to one another. The *polarization* of a light beam is defined as
the direction of vibration of the electric field.

To visualize this interlocking set of right angles, imagine

a beam of polarized light shining directly into your eye. If you could see them, the electric and magnetic fields would make a big X in your plane of vision. Along one arm of the X, the electric field is vibrating; along the other arm vibrates the light's magnetic component. Picture the electric arm of the X to be colored bright blue (electric blue?) while the magnetic arm is colored yellow (magnetic mustard?). The direction of the blue arm defines the polarization direction of the light beam. If the blue arm points in the vertical direction, for instance, what you are looking at is a beam of vertically polarized light.

Light is said to be totally polarized when its electric field vibrates in only one direction, and unpolarized when its electric field changes directions in an erratic and unpredictable manner. All other situations correspond to partial polarization. Direct sunlight is unpolarized, but scattered sunlight, skylight, for instance, is usually partially polarized to some extent.

To experience bee sight, it is best to begin by viewing light that is totally polarized, such as that obtained by looking at the sky through a sheet of Polaroid plastic or polarized sunglasses. Polaroid plastic is a transparent gray material that only passes light polarized in one particular direction—the direction of the plastic's optic axis.

As you look through the plastic at the sky, you will soon become aware of a polarization icon in the shape of a Maltese cross about five times as large as the full moon. One arm of the cross is blue, the other yellow. The cross has the visual character of an afterimage, and, like an afterimage, tends to fade away after a few seconds. To revive the icon, blink your eyes, shift your gaze, or rotate the plastic. When you turn the Polaroid plastic, the icon will rotate too, looking as though it were fastened to the plastic. Most people's first glimpse of the polarization icon (also called *Haidinger's brush*, after Austrian mineralogist Wilhelm Karl von Haidinger) occurs when looking through a rotating piece of Polaroid plastic. Much to their surprise, they suddenly see a big blue and yellow cross turning slowly against the sky.

After you are sure that you know what the polarization icon looks like in totally polarized light, try to see it in the partially polarized sky without the aid of the plastic. Best results are achieved at twilight against the background of a dark blue sky. When conditions are right, the sky sometimes appears to be covered with a latticework of yellow and blue crosses, an unforgettable sight.

The brightness of the polarization icon indicates how strongly the light is polarized. The icon is brightest for totally polarized light and fades to invisibility in unpolarized light. The arms of the Maltese cross point in the same direction as the vibrations of the light's electric and magnetic fields. The blue arm of the cross lies along the direction of electric vibration; the yellow arm indicates the light's magnetic direction. In the fringe science literature one runs across accounts of people who claim to be able to see colors around the poles of a magnet or "auras" around the human body. Whatever the merits of these claims, there is no doubt that ordinary people can, in a sense, perceive the magnetic and electric fields that constitute a beam of light. These fields appear to us as swatches of blue and yellow light. The physical source of the polarization icon has been attributed to special pigments in the eye or to the radial pattern of nerve fibers that overlay the retina, but the true cause of bee sight in humans is still obscure.

Once you have taught your eyes to experience bee sight, you will wonder how you ever missed seeing Professor Haidinger's wonderful multicolored crosses in the sky. Why was such a flagrant phenomenon—amounting to an entirely new human sense—overlooked by hundreds of generations of artists, explorers, and curious laymen until its relatively recent discovery (1846) by an obscure Austrian rock doctor? What other hidden human senses are awaiting discovery by alert sensory adventurers?

The Space of Voluntary Movement

"The will as brakes can't stop the will as motor for very long," said poet Robert Frost. "We're plainly made to go." And go we do, our mind skillfully coaxing and convulsing the body's 652 voluntary muscles into thousands of marvelous performances each day, from running a marathon to singing in the shower. Some of our muscles (from the Latin word for "little mouse") are under control of the will; others, such as the iris muscle in the eye, the muscles of the heart, or the tiny pilo-erector ("hair-raising") muscles that give us "goosebumps," receive orders from neural systems outside the range of conscious control. An important aspect of the human mind is its ability to move the human body willfully. How does it actually feel to exert conscious control over the movement of matter? What are the repertoire, range, and limit of our bodily powers?

The question of the limits of human performance is of great interest to human factors engineers who are designing man/machine interfaces for spaceships, airliners, and nuclear power plants. Psychologists, athletes, and dancers also work at defining and extending the body's limits. Human factors engineers measure the reach and grasp of human hands, the amount of force a human foot can exert on a pedal, the range of human reaction times, the "personal equation." How fast can a man run? How far can a woman jump? How far can she throw a ball, a javelin, a shotput? Every Olympic record of physical achievement is a psychological achievement as well, a record of the mind surging past its normal constraints.

The mind experiences its material boundaries in the form of a body schema—an ever-changing inner image of the posture, gait, expression, and appearance of the physical structure once called "the temple of the spirit." Part of our body image is constructed from information gathered via the external sense: we touch, smell, and hear the sounds of our own bodies in operation. Using our eyes, we catch sight of part of our body, and with mirrors see much more. My colleague, biophysicist-dancer Beverly Rubik, calls her mirror "an immediate multichannel biofeedback device."

Even with eyes closed I experience a strong sense of bodily presence: how my limbs are arranged, the position of each finger, where the tongue lies in my mouth. Much of our body schema comes from information garnered by the internal senses, notably the balance organ in the inner ear that tells the brain which direction is up, as well as proprioceptors (self-sensors) in the joints and muscles that directly inform the brain of the relative orientation of its body parts.

Like eyesight and hearing, the degree of body awareness varies widely from person to person. Some people are highly body-conscious; others body-blind. Psychological factors often distort our internal body maps, causing certain parts to appear larger or smaller than they really are. Drugs, fever, and delirium may radically alter our sense of the shape and size of the body. By far the greatest mismatch between body image and body fact is the phantom limb phenomenon, in which an amputated arm or leg still appears to be present. One man even claimed to be able to feel a wristwatch on his missing arm. Admiral Nelson, who lost his left arm in the Battle of Trafalgar, continued to feel its presence for the rest of his life. Nelson regarded the existence of his phantom limb as proof of the existence of a soul.

Attempts to create maps of human bodily possibilities have been few. Outside the field of human factors engineering, the work of Rudolf Laban and Ray Birdwhistell is particularly noteworthy.

In 1928 Laban published *Schrifttanz* ("Written Dance") in Vienna, introducing a new graphic system for mapping the possibilities of human movement. Laban visualized the dancer enclosed in a "kinesphere"—the space of all dance possibilities, the martial artist's "danger zone"—inside of which the joints of the body traced complicated paths. To Laban these paths resembled ribbons winding though a crystalline lattice. Because of the body's symmetry, a common "dance crystal" for these somatic meanderings was the icosohedron, the regular twenty-sided platonic solid. As developed by Laban and extended by his followers, Labanotation continues to be one of the most used body alphabets for the choreography of modern dance.

Ray Birdwhistell is an American anthropologist specializing in nonverbal communication. The goal of his "kinesics" project, initiated in the early 1950s, was to "develop a methodology which would exhaustively analyze the communicative behavior of the body." Birdwhistell found that Labanotation, originally developed for the annotation of dance movements, was not entirely suitable for the analysis of casual face-to-face communication. Birdwhistell began his system of kinesics by dividing the body into eight zones and inventing symbols to describe the motional possibilities of each zone. One of the advantages of a comprehensive movement map such as kinesics is to increase one's skill as an observer by drawing attention to normally ignored gestures such as subtle neck and shoulder movements. Birdwhistell, who claimed to be able to distinguish fifteen different degrees of eyelid closure, used his system to describe various styles of symptom presentation in Kentucky clinics, the American adolescent "courtship dance," body change when speaking a foreign language, and interruption strategies during therapy.

Some of the most awkward experiences that the new mind-link technology might be expected to engender will be the presentation to human consciousness of nonhuman body images. It may not be so difficult to take on the body schemas of dogs and cats, since their body plans are not dissimilar to our own, but to put on the body of a centipede or octopus may be a real challenge to our somatic imaginations. More difficult still will be the experience of mindlinking with microorganisms. Amoebas, for instance, have no fixed limbs at all, moving about, exuding pseudopods, and engulfing food particles by controlling the local viscosity of their cellular contents. What would it feel like to move around as a conscious gob of goo?

The possibility of assuming the body image of so formless a creature will bring new meaning to the phrase "Be all that you can be."

Thinking Space

The naked ape, pleased with what he does best, likes to call himself "the reasoning animal." Other beasts may possess keener senses, swifter movements, and, for all we know, deeper emotions, but there is no doubt that, compared to the other animals on this planet, man has developed his facility for rational thought to an almost grotesque degree. In the struggle for material existence, wisdom has become our main business.

At its core, the process of thinking depends on our ability to tell a good lie and stick with it. Metaphors Я Us. To think is to force one thing to "stand for" something that it is not, to substitute simple, tame, knowable, artificial concepts for some piece of the complex, wild, ultimately unknowable natural world. Much of the hard work of thinking has already been done for us by those anonymous ancestors who originated and shaped the earth's human languages. Language is surely one of our most useful tools of thought, giving conceptual prominence to certain things and processes, while relegating the unnamed and unnamable to conceptual oblivion.

Besides naming, other kinds of lying include reasoning by analogy or metaphor and the creation of legal, mathematical, personal, or social fictions such as money, limited liability corporations, the square root of minus one, enlightenment, and private property. Each word is a cultural enterprise, a public attempt to dissect the wordless world in one particular way. The usefulness of these verbal concepts should not blind us to the fact that a sudden insight, a change in fashion, or a new perspective may inspire other equally valid ways of construing the world riddle.

This aspect of thinking—the representation of one thing

by something it is not—is not restricted to mental processes. At the heart of all terrestrial biology lies the DNA code in which various triplets of organic bases have been made to "stand for" various amino acids. The most remarkable aspect of the DNA code is that the relationship between a triplet codon and its associated amino acid is not determined by chemistry or physics but is an essentially arbitrary assignment, analogous to the arbitrary association that humans make between a word and its referent.

Telling useful lies is only part of the enterprise of thinking. To be really useful these lies must be incorporated with explanatory intent into certain stories or games. A story is a narrative driven by essentially psychological motives. In this category fall myths, novels, theologies, parables, fables, and many kinds of modern therapies. A game is an organized system governed by impersonal rules rather than psychological motives. Examples of games include the monetary system, Euclidean geometry, rhetorical devices, rhyme schemes, Newtonian and quantum physics, real games of chance and skill, all maps, and all of mathematics.

One of the prime-time conceptual games of our era is called "mathematical logic," or the "truth game," under whose rules truthful new sentences can be mechanically generated from truthful old sentences. The rules of mathematical logic were first formulated in 1854 by Irish schoolteacher George Boole in a book he called *The Laws of Thought*.

The power of certain mathematical games (physics, chemistry, biology) to mirror the fine details of material existence faithfully is astonishing. In certain cases quantum physics makes predictions accurate to more than ten decimal places. Nobel Laureate Eugene Wigner refers to this magical match between human mathematics and nonhuman nature as "the unreasonable effectiveness of mathematics in the natural sciences." "This unreasonable effectiveness," concludes Wigner, "is a wonderful gift which we neither understand nor deserve."

Computers are symbol processors par excellence, subjecting symbols to the Boolean logic game much faster than any human mind can follow. However, the computer manipulates its symbols in a meaningless void. It does not distinguish between a new swim fin design, a ballistic missile trajectory, and a popular video game. The meaning of a computer's symbols is not understood or fixed by the computer but by its human programmers. In the world of thinking, humans are, among other things, the generators of meaning, and computers the unconscious executors of symbol games that bear (for humans alone) the burden of their meaning.

Besides meaning, what else do humans bring to the otherwise mechanical task of operating on symbols with game-driven rules? What, in other words, does thinking feel like?

Humans not only decide what the symbols stand for, they make up the rules as well. The same process of imagination that thought up Boolean logic has concocted other "logics" as well, suitable for calculating other kinds of truth. Part of conscious thinking is being aware not only of the symbols and the game in progress but also of alternative possibilities for changing the rules and extending the meaning of the symbols. Humans, along with other life-forms, are opportunistic, ready to change the rules of the game if it can afford them an advantage. Although we seem to possess strictly one-track conscious minds, the present moment of these minds seethes with myriad unrealized possibilities, the freedom to push thought processes in unprecedented new directions.

Besides giving meaning and imagination to game-driven thinking, humans also think in terms of stories, patterns of events driven by psychological motives rather than mechanical rules. Some stories make no sense unless you can imagine the emotions that the story's characters are experiencing. But what are human emotions?

Feeling Space

An emotion is a kind of psychological direction finder, orienting us toward pleasurable actions and away from painful ones—an internal compass of desire. Emotion helps draw our attention to particular things and events, making them stand out against a desire-neutral background, focusing our senses, giving clarity to our actions, and strengthening our memories.

Professor Robert Plutchik at Albert Einstein Medical Center in Philadelphia has devised the most systematic map of emotions to date. Plutchik distinguishes eight essential emotions that combine to produce all the others. Plutchik's primal passions come in pairs, each emotion matched with its complementary partner. A pure emotion combined with its complement results in an indecisive emotionally neutral experience.

Plutchik arrived at his eightfold catalog of feelings by collecting all the words for emotions in the English language, then arranging them in a systematic pattern, so that similar emotions were close together, dissimilar emotions far apart. In addition, this emotional positioning was guided by the hypothesis that each pure feeling is the subjective aspect of a particular biologically-based drive or need common to all animal life-forms from amoeba to human. An emotion is how a need feels. Thus a pure emotion should not only represent an unalloyed feeling but also correspond to the human version of some primal animal need.

The need to eat and the complementary urge to vomit (eat/excrete) are connected with the emotion of love and its complementary emotion, loathing.

The need for association with others and the complementary need to reintegrate oneself after severance from others (mate/separate) are connected with the emotion of joy and its complementary emotion, sorrow.

The need to defend oneself and the complementary need to retreat when defenseless (fight/flight) are related to the emotion of anger and the complementary emotion of fear.

The drive to explore the environment and the complementary need to maintain a home base (investigate/domesticate) correspond to the emotion of amazement and its complement, vigilance.

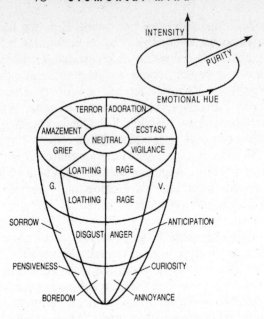

Inner-Space Emotion Map. Location of subjective feelings inside a three-dimensional emotional solid (after Plutchik).

Plutchik discovered that these eight basic passions could be arranged in a circle with complementary emotions opposite one another, to form a kind of emotional "color wheel." The pure emotions are arranged along the rim of this wheel, whose center corresponds to an emotionally neutral (white) state of mind. The fact that emotions can be experienced at many levels of intensity, ranging from mild to unbearable, adds a third dimension to this scheme, producing a three-dimensional space in which all passions might be mapped, a kind of "emotosphere," closely analogous to the three-dimensional color solid. Like the color solid's three axes of hue, saturation, and brightness, the emotion solid's dimensions might be labeled emotional hue (as in "hue and cry"), purity (of intent), and intensity (of feeling).

Although a considerable advance over previous crude

maps of feeling, Plutchik's emotosphere is still not as well developed as the color solid. The emotosphere is still largely qualitative and may not be exhaustive, and the mechanics of its mixed emotions have not yet been completely worked out.

Plutchik's emotion solid is a concrete example of the power of reason to organize even so unruly a field as the emotions, a successful example of clear thinking about feelings. The fact that human emotions seem to occupy a three-dimensional space similar to the color solid raises the intriguing question of whether the physical substrate of emotions might also be threefold. Could the spectrum of human emotions result from a trio of emotional "receptors" in the brain's limbic system, analogous to the trio of color receptors in the eye?

Memory

"Time," it has been said, "is Nature's way of keeping everything from happening at once." For humans, this same function is carried out by memory, without which we would dwell eternally in the present moment. Since the goal of many modern therapies and of certain Eastern religions is to live more fully in the present, futurologist John Holmdahl once playfully proposed an "Amnesia Foundation" for the popularization of memory loss as a shortcut to enlightenment.

The storage of memory traces seems to be composed of three separate mechanisms mediating short-, medium-, and long-term memorization. The length of short-term memory corresponds roughly to the time interval dubbed by William James, the "specious present," the time over which our experiences seem to be simultaneously available for mental manipulation. Medium-term memory lasts longer but can easily be disrupted, its contents forever forgotten. These temporary memories can be consolidated into long-term memory, capable of lasting a lifetime. The three kinds of memory may result from three different kinds of changes in the brain, corresponding roughly to electrical, chemical, and structural modifications.

Conventional computers consist of a small central processor plus massive amounts of memory storage space, each word of memory stored at a particular address where it can always be accessed. The brain differs from computers in that there seems to be no space at all allotted solely to memory. The brain's memorization facility seems to be diffused, in some ill-understood way, into the brain's sensory, motor, and emotional processing networks.

Human memories are not accessed by seeking a particular address, but instead evoked by association with other memories. The concept of "dog," for instance, may be linked to memories of particular dogs; to dog-related emotions of love or fear; to dog stories, shaggy or otherwise; to particular colors, smells, howls; to instances of canine communication, their connection with wolves, their domestication, the discovery of fire, and so forth. Every concept we can think of is connected to hundreds of others, in a type of memory structure called "associational" or "relational."

Early mind scientists from Aristotle to John Stuart Mill pictured the mind as a collection of concepts connected by associational links, much as atoms are connected by chemical bonds into large molecules. These "associationists" hoped that, once the rules of psychological connection were discovered, the structure of mental life could be simply understood as a kind of impersonal chemistry of ideas. An important research area in modern computer science—a modern echo of associational psychology—is the development of relational data bases—memory structures organized via a network of associations rather than by specific descriptors or by physical locations.

Higher Powers

One of the persistent beliefs about the mind, supported by much anecdotal evidence and some controversial laboratory findings, is that, under certain rare circumstances, the mind can exercise some of its powers independent of the body's mechanism, operating for a time free of the restrictions of the material brain. If we accept the possibility of extramaterial mental connections, then each of the previously discussed five powers of mind might also possess a paraphysical extension.

The extension of the senses into paraphysical realms has been called ESP, telepathy, dowsing, distant viewing, clairvoyance, and, when the senses break out of the present moment to access the future, precognition. Moving matter mentally, without the use of material muscles, has been dubbed psychokinesis, or PK for short.

Paraphysical cognition might include the spontaneous acquisition of ideas, inventions, theories, and proofs via extrasensory channels, deep intuitions arriving from outside the brain, telepathic contact with the Muses.

The word *telepathy* (from the Greek for "distant feeling") has been taken to mean direct sensing, but more means the extending of emotions beyond the brain, a kind of unmediated distant empathy with other conscious beings, mental "feelers" touching across vast distances.

Paraphysical extensions of memory would include accessing the fabled "Akashic Records," where the entire history of the universe is supposed to be stored; recalling events from "past lives"; or speaking in strange tongues (xenoglossia) that one has not learned in this life.

The question of whether any of these parapsychological powers exists or not is an important issue for consciousness research. One of the most confident predictions of materialist models of mind is that all such powers, every one, are purely fictitious. Dualist models of consciousness in principle permit direct mind-to-mind contact, but so far such models have failed to put limits on the range and the circumstances of unmediated mental connections. An adequate demonstration of one or more of these higher powers of mind would conclusively refute materialist mind models and might allow us to place realistic restrictions on current open-ended dualistic pictures of the mind/body connection.

Mind as we know it is characterized by sensations, voluntary bodily movements, memory, emotion, and cognition; it exists as a robust psychic unity (self) in a specious present roughly 3 seconds long. The self has the power at will to shift its attention (which resembles a serial data channel with a information capacity not exceeding 50 bits per second) over an immense field of possible activities and periodically dissolve and reconstitute itself in the familiar sleep/wake cycle. If other minds concurrently inhabit the body, this self is not aware of their existence. The material basis for the existence of our kind of mental life is the human brain, which has been called "the most complex arrangement of matter in the known universe."

consciousness from outside: a tour of the mind's mansion

Man has no body distinct from his soul
For that called body is a portion
Of soul
Discerned by the senses
The chief inlets of soul in our age.
—WILLIAM BLAKE

My body is that part of the world which can be altered by my thought. In the rest of the world my hypotheses cannot disturb the order of things.
—GEORGE CHRISTOPH LICHTENBERG

During the day Claire works as an entertainment critic for Universal Media Web (UMW); at night she often relaxes with a book from her archaic science-fiction collection. She especially enjoys reading stories about robots, is amused by old human fantasies about machines that obediently carry out human wishes. According to Claire, robot stories are the modern equivalent of Aladdin and his lamp; such stories voice the perennial human dream of exerting one-way power over nature. Magic lamps, however, always come with strings attached. People these days spend most of their time in the service of robotic needs—not because robots are stronger than people (many of them are, of course) but because robots have become indispensable to our way of life. Life without our electronic companions would seem impoverished, if not impossible. So perfectly do they serve us that we willingly became their slaves: some say that it started with the automobile.

Claire's favorite science-fiction story is Barrington J. Bayley's "Soul of the Robot." The hero, Jasperodus, a handsome young robot, finds himself superior to most humans in intelligence, but beset with a deep psychological problem. He experiences himself as possessing consciousness, a feeling no other robot seems to share, an experience from which robots are supposedly excluded for purely technical reasons. Attempting to resolve this dilemma, Jasperodus schools himself in the theory of advanced robotics, teaching himself enough cybernetic mathematics so that he can finally understand and assent to the proof that robots cannot ever possess consciousness. After this education he becomes doubly distressed. Not only does he feel an experience utterly alien to his kind, but he can now logically prove that such an experience cannot be happening. Thus not only is he a deviant robot but a damaged robot as well—one whose reasoning circuits keep making him jump to false conclusions.

After many adventures in the world of robots and men, Jasperodus returns to his human makers, an old man and woman near the end of their lives. The old man reveals a secret that explains Jasperodus's dilemma. It seems that, having no children of their own, they built Jasperodus to relieve their loneliness and placed inside his head elements of human consciousness taken from their own brains. Jasperodus discovers, to his delight, that he is no mere metal machine, but part of the human family.

What is so special about the human brain that allows it to wake every morning from matter's slumber and actually experience events in the world? What sort of outsides must a piece of matter put on before it can possess insides?

The ancient Egyptians thought so little of the brain that they discarded it—siphoning it out through the nostrils—before beginning their elaborate embalming procedure, while Ar-

istotle regarded it as a mere device to cool the blood. The seat
of consciousness has been located by many cultures in the
heart, in the liver, and even in the stomach, but we now
believe the brain to be the organ of mind. Three pounds of
oxygen-hungry meat, the brain is triple-wrapped in a series of
tough liquid-lined membranes, filled from within and bathed in
cerebrospinal fluid—cushioned and cherished by the body like
some precious embryo not yet come to term.

Brain in Embryo

Our brains developed in our mother's womb from a long hollow
structure called the neural tube. A few weeks after conception,
this tube was shaped like a long party balloon with three
prominent swellings near its closed end. These swellings de-
lineate the hollow chambers of the brain—the cerebral ventricles ("little stomachs")—filled with spinal fluid. The
substance of the brain develops around these embryonic chambers in three such different ways that anatomists use these
ventricles to separate our organ of mind conveniently into its
three major divisions.

The first chamber, blossoming explosively inside the pla-
centa like some lurid tropical flower, splits itself into two
subchambers to form the first and second cerebral ventricles,
around which the twin cortical hemispheres develop. The cerebral cortex, or *forebrain*, is by far the largest part of the
human brain, a thick convoluted sheet of neural tissue that
expands like yeast-rich bread inside the embryonic brainpan.
Its wildfire growth impeded by the skull, the double cortex
creases and bends back upon itself. Seeking more space it
grows forward; then up, over, and back; then forward once
more, completely enveloping the slower growing lower brain
chambers like some huge fleshy mushroom. Viewed from
above, the wrinkled forebrain resembles a huge walnut, divided down the middle by the great longitudinal fissure—the
cranial Grand Canyon. Viewed from the side, each cortical

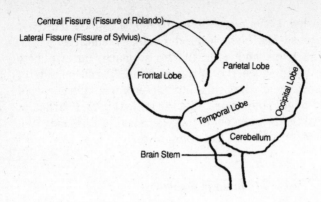

The Brain from Outside. The forebrain, or cerebral cortex, is divided into two hemispheres by a deep longitudinal fissure (not visible in this view). Each hemisphere is divided into four parts—the frontal, parietal, temporal, and occipital lobes—by the central fissure (or fissure of Rolando, a Sardinian physician) and the lateral fissure (or fissure of Sylvius, a Dutch anatomist). The forebrain completely covers the midbrain structures like an umbrella, exposing only parts of the hindbrain: the cerebellum ("little brain") and some of the brainstem.

hemisphere resembles a crumpled boxing glove, the front, middle, and back of the glove corresponding to the brain's frontal, parietal (Latin for "wall"), and occipital ("back of the head") lobes, while the boxing glove's thumb corresponds to the brain's temporal lobe.

Around the second neural chamber, the more modest growth of the *midbrain* structures take place. Stretched across the top of this central cavity is a pair of C-shaped for-nixes ("city arches," beneath which the Roman *fornicatrix* met her horny clientele), a part of the brain's *limbic system*, believed to be the material basis for the emotional life. On each side of this central ventricle lie the left and right *thalamuses* ("bridal chambers"), while the floor of this ventricle is called the *hypothalamus* ("underneath the bride's room"). Sprouting out from the region of the third ventricle, like so many exotic fruits, are the olfactory lobes, the pituitary and mammillary bodies, the pineal gland (which Descartes guessed to be the seat of the soul), and eight little bumps on the back of the brain that early anatomists called, in rhyming Latin, knees *(geniculi)* and hills *(colliculi)*.

The grotesquely swollen cerebral cortex covers the midbrain like an umbrella. The handle of this umbrella is the brain stem, or "hindbrain." Relatively undeveloped, except for the cauliflowerlike efflorescence of the cerebellum in the back of the brain, it resembles a thick-walled hollow tube—with the fourth ventricle as its core—bulging twice in front to form two lumps, the pons ("bridge") and the euphonious medulla oblongata ("elongated core").

Contemporary essayist Ihab Hassan describes the brain this way:

> The brain is not yet whole or one. Like a divided flower, never exposed to the sun, it grows from an ancient stem that controls both heart and lungs. On each side, cerebellum, thalamus, and limbic system twice grasp this stem. Our muscles, our senses, our rages and fears and loves, in this double fistful of old matter stir about. The great new cortex envelops the whole, grey petals and convolutions, where will, reason, and memory strive to shape all into mind.

The brain is supplied with blood mainly by the twin carotid arteries, its wastes carried away by the jugular and other veins. These major conduits branch into an intricate network that sprawls across and into the cortex supplying the brain's metabolic needs: food for thought. The brain burns glucose (a right-handed sugar) and oxygen like any other body organ. Its power requirement is about 20 watts, much less than your 100-watt reading lamp. The body at rest consumes about 80 watts (basal metabolism). The brainy 2 percent of the body's bulk grabs 25 percent of the body's energy. The job of minding the body is hard work. The brain spends the lion's share of the body's energy budget.

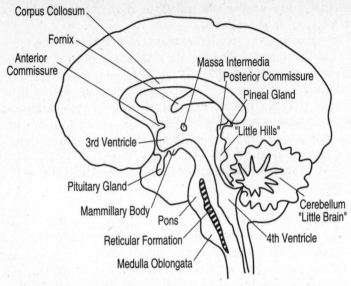

Inside the Brain. A section of the brain along the longitudinal fissure shows some of the parts normally covered by the cerebral cortex. Four neural "cables" join the left and right hemispheres: the corpus collusum, anterior and posterior commissures, and the massa intermedia connecting the two thalamuses. The fornix and mammillary body, along with the septum, hippocampus, and amygdala (not shown), are part of the brain's limbic system, which mediates emotion. The colliculi ("little hills"), along with the geniculi ("little knees"), act as relay stations for the visual and other sensory systems. The pituitary gland (Latin for "spit" or "slime"—it was once thought to be the source of nasal mucus) produces many of the body's regulatory hormones. The function of the little pineal gland, which Descartes guessed to be the seat of the soul, is unknown.

Powers of Mind

It is of some interest to compare the power consumption of the brain in various states of consciousness. The following list (after Seymour Kety) compares the power consumed by the brain in various mental states as measured by cerebral oxygen consumption, a quantity Kety calls "brainwork."

Mental arithmetic: 102 percent
LSD state: 101 percent

Schizophrenic state: 100 percent
Normal awake state: 100 percent
Sleep state: 97 percent
General anesthesia: 64 percent
Insulin coma: 58 percent

Within the errors of measurement, the first five states of consciousness listed here use about the same brainpower. The brain works no harder doing complex mental arithmetic than it does sleeping. In fact, says Kety, the beating of the heart is a better index of mental effort than brainwork. Only in states of deep anesthesia or coma does the brain's power consumption drop significantly.

If we take seriously the 3 percent difference between the waking and sleeping states, we can roughly estimate the power necessary to maintain ordinary waking awareness. It amounts to little more than half a watt—about as much power as that consumed by a pocket flashlight. Could this tiny power difference between the conscious and unconscious states be interpreted as the electric equivalent of "willpower"?

In the middle of the nineteenth century, physicists in many countries formulated the principle of conservation of energy. Energy occurs in many forms—chemical, mechanical, gravitational, thermal, and electrical, for example. In 1905 Einstein showed that mass itself is a form of congealed energy. The law of energy conservation states that in a closed system, the amount of energy must remain constant. Energy may change its form—chemical into mechanical, as in muscle; light into electrical, as in the eye. But no energy is ever lost or created in such transactions. Despite the revolutions in science wrought by relativity and quantum theory, the energy conservation principle has remained intact.

If consciousness is a new form of energy, we should be able to measure a number in the brains of conscious creatures that represents its energetic equivalent: so much experiential consciousness (measured in experienced bits per second) is equal to so much electrical power (measured in watts). How many thoughts in a watt? Kety's measurements on the energy consumption of the brain combined with the consciousness data rates discussed in the previous chapter give us some rough idea about how large this (presently unknown) number might be.

Kety's approach is crude—comparable to examining the operation of a TV set by reading the electric bills—but it at least gives us some approximate idea of the amount of electrical work necessary to produce the familiar psychic activity we call ordinary awareness.

Looking into the Brain

Since the brain has now been identified as the organ of consciousness, mind scientists are particularly eager to find ways of observing this organ as it goes about its business, rather than just eying it cold and lifeless on the anatomist's dissection slab.

Certain extensions of the surgeon's art have increased our knowledge of the living brain, such as Wilder Penfield's direct stimulation of the exposed cortex of conscious patients and the excitation of human pleasure and pain centers deep in the brain by James Olds and his colleagues. However, surgical investigation of the living brain can be justified only in unusual situations. We need less invasive tools than the surgeon's blade to satisfy our intense curiosity concerning the day-to-day operation of the human organ of mind.

Alfred Nobel, the inventor of dynamite, set up the Nobel Prizes to recognize outstanding achievements in science, literature, and world peace. The first Nobel Prize in physics was awarded to Wilhelm Roentgen for his discovery of X rays. X rays, a form of electromagnetic radiation 10,000 times more energetic than ordinary light, have been of invaluable aid to medical science, producing the familiar shadowgraphic pictures of the insides of the living body. The old-fashioned X-ray photo has been recently supplemented by a modern technique involving a movable X-ray source and solid-state detector arrays. The information from these detectors is collected by a computer and assembled into a three-dimensional image of the brain or other body part, a technique called CAT scan (computerized axial tomography). Because flesh is relatively transparent to these rays, X rays are more suited to visualizing bone structures than to studying soft tissue like the brain. To make your intestines more opaque to X rays, doctors will commonly treat you to a barium (heavy metal) enema.

A recently developed technique called NMR imaging (for nuclear magnetic resonance) or MRI (magnetic resonance imaging) is able to produce detailed pictures of soft tissues that complement the bone scans produced by X rays. Furthermore, the MRI device probes the body with a combination of magnetic fields and radio waves, both of which are harmless, as far as we know, compared to the somewhat detrimental effects of X rays. The MRI device works by provoking the hydrogen atoms in the body to give off weak radio signals and then constructs a three-dimensional map of the intensity and location of these atomic radio stations. Since most of the body's soft tissues are composed of water, which is two-thirds hydrogen, this procedure produces a remarkably detailed picture of the body's fleshy insides, including the brain. The MRI device produces a static image of the brain: other techniques are used to map the brain's ongoing activity.

In the BEAM (brain electrical activity mapping) technique, multiple electrodes are attached to the scalp and the voltage of each electrode sent to a computer, which displays the shifting electrical patterns on a color screen. BEAM is a kind of real-time cerebral weather map, revealing perhaps the location of electrical "brainstorms" on the cortical surface. However, electrical activity at the surface of the head only dimly reflects the complex activity inside.

Since electrical activity in the brain stem is only indirectly reflected in scalp voltages, large changes in consciousness—from coma to wakefulness, for instance—can occur without correspondingly large changes in scalp electricity. The brain seems almost purposely designed, like the shielded cables that feed your VCR, to prevent electrical signals from leaking to the outside world.

The brain is immersed in a relatively conductive fluid and surrounded by moist conductive supportive tissue, analogous to the conductive metal ground sheath that electrically shields your TV cable. Then the brain is covered with a thin insulating layer—the bony skull—analogous to the TV cable's protective rubber coating. Then, for good measure, a second conductive membrane, the hair-covered scalp, is stretched over the skull. Any electrical signals that manage to force their way through the brain's multiple electrical barriers must be very robust and certainly not representative of the subtle electrical changes going on deep inside.

Although the head is relatively opaque to electrical signals, it is completely transparent to magnetism. Since every electrical signal in the brain also produces a weak magnetic field, a magnetic brain wave sensor could, in principle, provide an undistorted picture of real-time deep electrical activity in the brain. Crude pictures of the brain's magnetic activity have been achieved by SQUID (superconducting quantum interference device) magnetometers, but further improvement of magnetic activity mapping in the brain is hampered by the intrinsic weakness of magnetic brain signals (brain magnetism is 100 million times weaker than the earth's magnetic field), and the large effective size of the superconducting detectors, which must be enclosed in big vacuum flasks cooled down to near absolute zero. One immediate consequence of the hoped-for room-temperature superconductor would be a vastly improved ability to picture magnetic activity deep inside the living brain.

Another method of visualizing the brain's inner activity is the PET (positron emission tomography) technique. The PET technique introduces a short-half-life radioactive sugar into the bloodstream. This sugar is incorporated preferentially in those parts of the brain with the highest metabolic activity, the "hardest-working" brain centers. The sugar signals its presence in the brain by emitting a positron (a tiny bit of antimatter), which explodes on contact with ordinary matter to produce two powerful gamma rays, a type of radiation for which the brain is almost transparent. Gamma-ray detectors arrayed around the head pick up the two rays, and a computer traces their paths back to their place of common origin deep inside the skull. These gamma-ray pairs act as pointers allowing the computer to display the shifting pattern of the brain's sugar metabolism as a three-dimensional color TV image. PET is a dynamic version of Kety's brainwork measurements, not only recording overall changes in the brain's metabolism but actively picturing the changing distribution of brainwork among the various brain centers as the brain's owner carries out a variety of mental tasks.

After years of probing the dead brain with the anatomist's scalpel and peering at dissected brain cells with optical and electron microscopes, we are just beginning to look deeper into the brain at work and play through the clever use of radio waves, gamma rays, as well as the brain's own electrical and magnetic impulses.

How Does the Brain Work?

The human brain has been described as the most complex object in the universe. Certainly a lot goes on in this warm fist-sized ball of meat. Various exotic fluids pour, soak, and trickle through its channels and crevices. A veritable drugstore of chemical substances is synthesized there, put to strange uses, then broken down and recycled for further use. Legions of brain cells are born (in the early months of life), connect up to other cells, and carry out their mysterious cellular tasks in various neural communities before they die. Trillions of electric signals travel through the brain's wet electrical networks, each impulse inducing a weak electrical and magnetic field that races across the cranium at the speed of light. Torrents of electrically charged ions escape through suddenly opened cellular gates only to be captured one by one and sequestered again inside a brain cell. In addition, if the dualists are right, certain special brain processes act in unknown ways to send and receive messages from the spirit world. With so much activity going on all at once, it is difficult to tell which brain functions are important, which irrelevant, for producing the phenomenon known as ordinary awareness.

Because the brain takes in a disproportionate amount of the body's blood, Aristotle may be forgiven for supposing it to be a blood cooler. Even after the brain was recognized as the organ of thought, hydraulic metaphors continued to be popular to explain its operation.

Early Greek and medieval physicians pictured the body as primarily a network of tubes, valves, pumps, and reservoirs through which coursed various liquid "humors," or vital fluids. These included the tangible fluids, such as bile, blood, phlegm, semen, and lymph, as well as the more rarified fluids: vital and animal "spirits." These spirits were connected in some way with the action of mind, with sensations, mentation, and voluntary activity. They accumulated in the ventricles of the brain, were in some unknown fashion responsible for psychological functioning, and moved the body's limbs by flowing through hollow nerve fibers to inflate some muscles and deflate others, much the same way that high-pressure oil expands the hydraulic pistons of a bulldozer or airplane landing gear. Today we believe that the only "muscle" in the body that works by fluid inflation is the blood-expandable penis.

One reason for Descartes's choice of the pineal gland as the site of the seat of the soul was that this gland is located near the intersection of the three major ventricles of the brain, an opportune location for directing the flow of vital spirits in this organ of mind.

Today we consider the hydraulic movements of blood and cerebrospinal fluid in the brain as largely irrelevant to the operations of mind. Electrical metaphors are currently in vogue; the brain is regarded now as a kind of electrochemical computer made of meat. We know, for instance, that the nerves are not hollow tubes for the transport of vital fluids but more like telephone transmission lines carrying electrical pulses into the brain from the sense organs, and out to the muscles. What happens to these pulses inside the brain is less clear. Unlike a silicon-based computer, the brain's central processor is not entirely electrical, but involves complicated chemical processes, and perhaps mechanical operations as well. The brain, after all, is not made up of inert plastic chips; it is a biological community made of billions of living beings.

The brain consists mainly of two types of living cells, the long stringy neurons (Greek for "bowstring") and the compact glial ("glue") cells. The glial cells are at present assigned only a supporting role in the physiological processes of mind, nursemaids to the more important neurons.

Each neuron is an enormously elongated cell with a foliage of input lines (dendrites) and a single output line (axon). Axons may be as much as a meter in length, for instance, the neuron that runs from spinal column to foot, making the nerve cells by far the largest cells in the body. The narrow communication threads of the neuron are less than a hair's diameter, but they are commonly clustered together into bundles of thousands of individual nerve fibers to form large ropelike nerve tracts of "white matter." The cell body out of which the dendrites and axons grow like the arms of an octopus contains the nucleus of the cell—the locus of its genetic code—as well as the blood-fed metabolic machinery that keeps the cell alive and electrically active. The cell bodies are slightly darker than the translucent neuronal "cables"; associations of cell bodies form the gray matter of the brain and spinal column.

Each cell in the body, not just its neurons, is a tiny electrical battery, powered by a difference in concentration of sodium and potassium ions across the cell membrane. In most kinds of cells, this battery function plays no known role, but in the nerve cell, changes in local battery voltage are used to transmit electric signals along the cell's dendrites and axons. When a portion of the nerve-cell battery is discharged by some external influence—electrical, chemical, or mechanical—it quickly returns to its normal voltage. However, if the discharge is intense or prolonged, the nerve cell does not bounce back. Instead cell-battery discharge is triggered in neighboring regions of the cell membrane, which in turn triggers more discharge farther away. A self-sustaining wave of electrical discharge—the "nerve impulse"—begins to travel along the nerve membrane. In the wake of the discharge wave, the neuron slowly returns to its original voltage. This ability of the nerve cell to sustain a localized traveling wave of electrical discharge is responsible for communication between sense organs and the brain, between the brain and its muscles, and for much, but not all, of the brain's computational activity.

It was once believed that the nervous system was a continuous network—every cell tightly connected to its neighbor. However, it was soon discovered that nerve cells never actually meet: instead at points of contact they are separated by a tiny gap called the "synapse," a gap too large to be bridged by the weak electrical signal produced by membrane discharge. Instead of facilitating direct electric transmission, each synapse is a kind of neural customhouse where electricity is changed into a chemical currency. An electrical discharge on one side of the gap induces its nerve to emit preformed packets of a certain chemical—called the "transmitter substance" —that quickly diffuses across the watery synaptic moat separating the two nerve cells. Upon arrival at the second nerve cell the chemical either tends to induce discharge (excitatory synapse) or repress it (inhibitory synapse).

Several dozen different transmitter substances are known to exist, seemingly segregated into particular nerve networks, the dopamine-mediated network in the midbrain, for instance, or the serotonin network in the brain stem. Although all the nerve cells of the brain are essentially alike, the synaptic "customhouses" that connect them use many different chemical currencies. We have begun to trace the distribution of common-currency synapses and may soon possess a color-coded atlas of the brain's many interconnected chemical bailiwicks.

The electrical signals produced by these chemicals at the

synapses travel along the dendrites to the cell body, where they add together to produce a composite signal. If this composite signal rises above a certain threshold voltage, the cell's axon fires, sending a nerve impulse along its length, to synapse with another neuron, or to trigger a muscle contraction. If the composite signal remains below threshold, the axon does not fire and the nerve cell does not participate in the web of communications going on around it. A neuron with n synapses resembles a kind of computer element called the n-input AND gate. The AND gate fires only when all of its inputs are stimulated at once. The neuron differs from the AND gate by being more complex and less reliable. The same electrical input to the AND gate always results in the same response. Not so for the neuron. Many other factors besides electrical input determine whether a synapse will trigger its adjacent nerve cell: for instance, the temperature of the cell; its previous history; the electrical field produced by adjacent neurons; the presence or absence of certain chemicals called neuropeptides, which drift between cells as "slow chemical messengers"; and perhaps, some say, the quantum uncertainty mandated by Heisenberg's uncertainty principle.

Because the synapse is subject to so many factors besides the electrical input, computer scientist Ernest W. Kent refers to the nerve cell as a "MAYBE gate." Is the unreliability of neurons an unavoidable drawback of a computer that must be constructed from living beings, or is this unreliability necessary for the kind of "computations" the brain must carry out in order to be conscious? It might be advantageous for a being living in an uncertain world to possess a somewhat unreliable decision mechanism: two wrongs may sometimes make a right.

Where Is Consciousness?

The search for the location of human awareness in the brain takes two directions: the elimination of certain brain sites by elucidating their nonconscious bodily functions and the identification of those brain sites that are occupied in conscious-ness-related functions, such as shifting attention, producing voluntary movement, and modulating the sleep/wake cycle. To narrow the search for mind sites, it is just as useful to know where consciousness is certainly not present in the brain as to know where it might be located.

The brain may be roughly envisioned as a series of three concentric shells: the cortex, various subcortical structures, and the centrally located thalamus, all straddling the top of the brain stem, which itself is a specialized extension of the spinal cord. The brain's three shells are split down the middle along the great longitudinal fissure into left and right cortical hemispheres, left and right subcortical structures, and left and right thalamuses. The brain stem is not physically divided in two although it is symmetric along the left-right axis.

The cerebral cortex (Latin for "rind") is a crumpled layer of gray matter the thickness of an orange peel with the con-sistency of tapioca pudding. The wrinkled cortex makes up seven-tenths of the entire nervous system, containing perhaps 8 billion nerve cells interconnected by almost 1 million miles of nerve fibers. Without the skull to contain it, the custardlike cortex could not support its own weight. The cortical material is so soft that brain surgeons often use, instead of knives, tiny vacuum straws called "slurpers" to cut into the flesh of the forebrain. By means of electrical stimulation of the exposed brain surface, the functions of the cortex have been largely mapped out and the cortex divided into sensory, motor, and "association" areas.

Each of the five senses has a region on the cortex where signals from their sense organs converge. The visual cortex occupies most of the occipital lobe. The sense of hearing is lodged in the crease where the thumblike temporal lobe joins the parietal lobe, while the sense of smell resides on the lower inside surface of the temporal lobe, a portion called the uncus ("hook") that bends around the brain stem.

From all parts of the body, touch receptors send tactile messages to the center of the back, up the spinal column into the brain stem, through thalamic relay centers to the sensory cortex, a ribbon of tissue that lies directly behind the central fissure of Sylvius. Each section of this sensory ribbon is associated with the tactile sensation from a single body part so that the entire body surface is mapped touchwise onto a narrow band of neural tissue. This cortical mapping allocates more cortical space to sensitive organs such as lips and hands. This distorted mapping of body parts to brain tissue—beginning with the feet, toes, and genitals inside the longitudinal fissure and ending with the tongue and throat in the lateral (Rolandic) fissure—is called the "sensory homunculus." The sense of taste is located in the facial area of the sensory homunculus: as far as the brain is concerned, our taste sense seems to be treated as another form of touch.

Across the central fissure from the sensory homunculus lies a corresponding "motor homunculus," where the parts of the body that can be moved voluntarily are mapped onto a narrow ribbon of cortical tissue—starting with the feet inside the longitudinal fissure and ending with the tongue near the lateral (Rolandic) fissure. Slightly forward of the tongue site on the dominant hemisphere lies Broca's area, concerned with the motor movements involved with speech. Forward of the motor homunculus lie two cortical regions also concerned with muscle control, the supplementary motor area and the premotor cortex.

The cortex is believed to be the site of the brain's memory function, but unlike computers, which dedicate a large fraction of their space entirely to memory storage, the brain seems to have no area devoted explicitly to memory. Instead it is believed that memory is somehow "distributed" throughout the brain's sensory and motor circuits. The claim of certain piano players that their musical ability resides in their hands not in their heads may not be so farfetched.

Those parts of the cortex not devoted to sensory or motor tasks are called, for lack of a better name, the "association areas" and are thought to be devoted to intellectual tasks, making sense out of the world as well as "making the world" out of sense—constructing a logically consistent picture of the outside world, of the body's place in that world, and of the "inside world" of the mind. When muscle movements, sensory experiences, or past memories are induced by direct electrical stimulation of the cortex, these experiences are always felt to come from outside, not to be initiated by the self. Although the cortex seems to be responsible for much of the contents of consciousness, no part of the cortex has yet been found that mediates the experience of consciousness itself.

Beneath the cortex lies a second shell of subcortical structures that surround the third ventricle: the basal ganglia, the limbic system, and the hippocampus.

The basal ganglia are large tadpole-shaped bodies arching their backs beneath the cortical mantle. They are involved in shaping voluntary muscle movement and are the locus of Parkinson's disease.

The limbic ("border") system consists of the mammillary bodies, the C-shaped fornix, the hippocampus, the amygdala ("almond"), and other minor structures. Stimulation of the limbic structures induces feelings of pleasure, rage, anxiety, agitation, and cheerfulness. This portion of the brain is evidently responsible in some way for the emotional content of our experiences.

The hippocampus ("seahorse"), a twisted structure lying along the inside margin of the temporal lobe, in addition to being a part of the limbic system, seems to be involved in the consolidation of memory traces. If the hippocampus is excised or damaged, no long-term memories can be formed.

The paired thalamuses form the brain's innermost shell. Through these organs all of the body's sensory signals (with the exception of smell, which is channeled directly to the limbic system) and most of the body's motor signals pass between cortex and spinal column. They are the brain's central relay station—the master electronic switchboard—an anatomically correct place for some modern Descartes to locate consciousness now that electric metaphors have replaced hydraulic ones for explaining what the brain does.

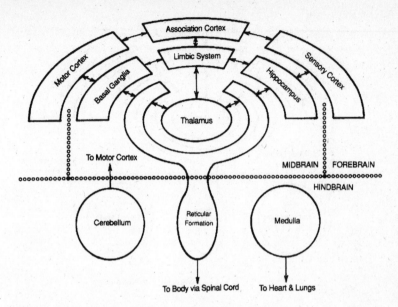

Three-shell model of brain structure, showing the central location of the thalamus and reticular formation among the neural networks that electrify the body.

The two thalamuses sprout out of the brain stem, an unpaired but symmetric extension of the spinal cord, which consists of the pons, the medulla, and the cauliflowerlike cerebellum. The cerebellum ("little brain") seems to act as a motion computer that handles posture as well as certain aspects of voluntary motion. The medulla contains timing circuits that regulate the operations of the heart and lungs. Both the pons and the medulla contain the roots of the "cranial nerves," special nerve centers that subserve the sensory/motor functions of the face, the most highly structured human body part. Deep inside the brain stem lies a diffuse network of neurons called the reticular formation.

The reticular formation satisfies what might be called the "Descartes criteria" for a likely site for consciousness, namely that, to explain our unity of mind, the consciousness organ should be unpaired, and to fulfill its central executive function

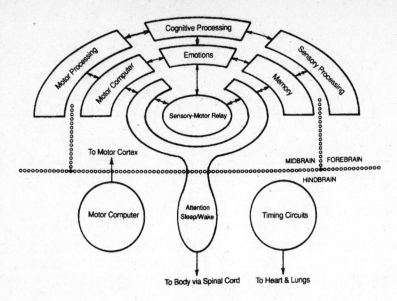

Three-shell model of brain function showing the central role of the thalamus and reticular formation in modulating the sensory, motor, and mental (attention) processes that enliven the human body/mind.

it should be centrally located. Like the thalamus, the reticular formation is centrally located and straddles the major sensory and motor pathways going to and from the brain. Unlike the thalamus, the reticular formation is unpaired and possesses additional properties that make it a more likely candidate than the thalamus for the location of the brain's consciousness mechanism.

The Reticular Formation

The reticular formation ("reform" for short) occupies the central portion of the spinal cord and extends from the base of the spine, through the brain stem, and up into the thalamus. The top of the reform consists of a thin sheet of gray matter called the "reticular complex" covering like a veil part of the rear surface of the thalamus. As its name implies, the reticular formation is a diffuse netlike arrangement of neurons that extends its arms out across the brain stem from a central core bordering the fourth ventricle and spinal canal. The reticular formation contains about 1/1000 as many neurons as the cerebral cortex, but all major sensory and motor pathways must pass through this diffuse neuronal thicket on their way to and from the brain.

The structure of the reticular formation has been compared to a stack of fuzzy poker chips stacked along the spinal column. Kilmer and his colleagues at MIT have described the function of the reticular formation stack as

> the nervous center which integrates the complex of sensory-motor and autonomic-nervous relations so as to permit an organism to function as a unit instead of a mere collection of organs. Its primary job is to commit the organism to one or another of about 16 gross modes of behavior—i.e., run, fight, sleep, speak—as a function of the nerve impulses that have played in upon it during the last fraction of a second.

Thus the reticular formation seems to perform at least one function that we attribute to consciousness, making the moment-to-moment decisions about what the whole body should do with itself.

Besides deciding what to do in a particular situation, exercising what might be called the "motor will," the reticular formation seems to be involved in choosing what aspects of sensory input the brain pays attention to, the exercise of "sensory will." Nobel laureate Francis Crick recently proposed that the portion of the reticular formation that surrounds the thalamus performs a "searchlight" function by isolating which parts of the sensory information flowing through that central relay station will be enhanced, which suppressed. "If the thalamus is the gateway to the cortex," says Crick, "the reticular complex might be described as the guardian of the gateway." The mechanisms for other kinds of selective attention, conscious logical operations ("cognitive will") or conscious activation of old memory traces ("recall will"), have not been elucidated, but since the reticular formation is already implicated in the selective activation of sensory and motor cortex, it is not too farfetched to imagine that this central brain stem organ might also selectively activate cortical association areas to initiate conscious reasoning or portions of the cortex that subsume the storage of memory traces.

In addition to regulating various kinds of attention, the reticular formation and associated brain stem structures are responsible for the sleep/wake cycle in humans and other animals. The consciousness function is turned off in sleep and reinstated once again in the waking state. In this context the sleep state may be considered just one of the sixteen-odd classes of behavior that the reticular formation can commit the entire organism to carrying out. But sleep, unlike other forms of behavior, has a more immediate relationship to the mind/body problem since it seems to involve the temporary abolition of mind.

One of the most obvious facts about consciousness is how easy it is to lose it. Lack of oxygen, lack of sugar (insulin coma), or damage to the brain stem can disrupt conscious awareness even though the body's other systems remain operational. One of the earliest experiments on the mind/body problem—first carried out, no doubt, by some anonymous caveman—was the observation that a blow on the head often leads to loss of consciousness whereas a painful blow to the foot leaves consciousness intact. The explanation of the relationship between a caveman's club to the head and the temporary disruption of his victim's inner life is still forthcoming.

One of the most ingenious methods for investigating the coma-inducing mechanism of cerebral concussion was invented by Holbourn, who constructed a model of the brain out of gelatine enclosed in a hard wax skull. Holbourn attempted to relate the location and severity of breaks in the gelatine with different kinds of blows to his simulated skull. Two mechanisms leading to unconsciousness were identified, both involving damage to the brain stem. Twisting of the head (whiplash) caused central brain stem structures to be sheared by a kind of whirlpool action, as in a food blender. Second, compression of the braincase forced the soft brain material to be extruded like toothpaste out the hole at the bottom of the skull, damaging the brain stem in this vicinity. Both of Holbourn's observations implicate the brain stem as that part of the brain essential to maintaining the experience of conscious awareness.

In addition to being subject to damage during concussion, the reticular formation is the site of action of drugs that modify consciousness itself rather than its contents. Amphetamines and barbiturates, chemicals that increase or suppress our sense of inner presence or "psychic energy," act primarily on the brain stem. The reticular formation is also the location where general anesthetics exert their effect. These are chemicals (generally small molecules such as nitrous oxide) that quench consciousness completely. Understanding the physical basis for anesthetic action is an important ongoing area of awareness research.

All of the neurons that utilize the chemical serotonin as a transmitter substance are located in the reticular formation. Recently LSD has been identified as a molecule that directly competes with serotonin for the occupation of synaptic receptor sites. In addition to heightening perceptions and emotional experiences, LSD also alters the perceived nature of personal identity. Unlike other drugs, which modify the contents of consciousness leaving identity intact, LSD and similar "psychedelics" seem to work on consciousness itself, to modify centrally the self-aware core of our being.

In his long career as a neurosurgeon, Wilder Penfield had the opportunity to stimulate the cortexes of more than 1000 conscious patients electrically and to listen to the subjective reports about what such stimulation feels like. Although he could radically modify the contents of consciousness with his electrodes, Penfield never once was able to touch the central core of the patient's being, leading him to speculate that the source of mind was not in the cortex, but somewhere else, perhaps in the brain stem, where his electrodes could not reach, or even (the extreme dualist position) entirely outside the body.

These arguments for the role of the reticular formation as the organ of conscious awareness are not conclusive. For example, the switch of a TV set turns off the set, just as damage to the brain stem turns off consciousness, but the operation of the on/off switch does little to explain how a TV set works. A model of awareness in which consciousness is produced solely in the reticular formation may be too naive. A better picture might involve a certain inseparable collaboration between cortex and brain stem. One stumbling block for a wholly reticular model of consciousness are the split-brain experiments of Roger Sperry and others, in which information presented to the right side of the brain can be processed and acted on without ever entering conscious awareness; consciousness in these split brains only seems to have access to information stored in the dominant (usually left) hemisphere. Since both halves of the brain remain linked at the level of the brain stem, one is at a loss to explain why brain-stem-induced consciousness only seems to flow into the dominant hemisphere.

The choice of the reticular formation as the seat of the soul is not unanimous. Brain researchers are still uncertain as to the location of the consciousness mechanism in the brain. Some locate it in the cortex, some in the brain stem, others in the interplay between cortex and stem. I side here with those who associate consciousness with the reticular formation. Other brain parts contribute to the detailed contents of consciousness, I believe, but are not essential for its presence.

This rough guess about the function of the brain locates consciousness near the junction of the midbrain and the hindbrain. Here is where the central executive dwells who selects, chooses, and above all experiences some of the activities carried out by the other brain structures. Here is where our search for the secret of human consciousness rightly begins. What is so special about this nervous tangle—about 10 million neurons, the population of Tokyo—that fits it for such an important role? How does the reticular formation manage to turn meat into mind?

Brain research in the past has been guided by metaphors borrowed from the prevailing technology of the times. Thus we have witnessed hydraulic, telegraphic, switchboard, and holographic models of the brain. More recently the brain has been compared to a computer.

Most present computers are digital (yes/no data only) and serial (performing only one operation at a time). The brain, on the other hand, seems to be a hybrid (both yes/no and graded data) and parallel (many simultaneous operations) machine. The theory of large, hybrid parallel machines is in an embryonic state and has not yet contributed much to brain research. This is an area of great ignorance and of correspondingly great opportunity for fruitful research.

Though brains may differ from computers in many details, there are some functional similarities between the two. Present-day digital computers consist of a central processing unit (CPU) that handles the actual computations and sequence control; various kinds of input/output devices for communicating with the world outside the computer; and memory units to store both programs and data. The heart of a computer lies in its CPU. Memory and input/output devices are considered "peripherals" to the central processor.

In our brain model, the reticular formation plays the part of a computer's CPU; the sensory/motor cortex, along with basal ganglia and cerebellum, handles input/output routines. Memory in the brain is not segregated into one particular location as in a computer but is distributed in some unknown way among the brain's input/output machinery. Since present computers possess (as far as we know) no internal experiences, there is a natural limit to our analogy.

The enormous elaboration of cortical area that distinguishes humans from the other animals has not been matched by a corresponding growth in the complexity of the reticular formation. The evolution of the cerebral computer has been achieved not by growing a bigger central processor but by acquiring more powerful peripherals, most important for humans, a facility for spoken language. To prevent the reticular CPU from being overwhelmed by the organism's enriched sensory/motor capacity, these peripherals perform a great deal of autonomous preprocessing. In the jargon of the computer programmer, our senses and our muscles behave like "smart terminals."

Laid down in the cortex are our language, world map (including our notions concerning the brain and consciousness), and personal memories. Awareness and control seem to be lodged in the reticular formation. For the quality of experience that we have come to regard as normal awareness, the cortex is absolutely essential—especially the memory function, which confers continuity to our presence. Without the cortex, we might experience a bare-bones sort of awareness, but it would not be human awareness.

For all of its importance in establishing the quality of our consciousness, the cortex seems to be an essentially mechanical structure. It does not produce consciousness, though it does substantially augment it. The mysterious physics of ordinary awareness and its possible nonphysical extensions lies coded into the structure of the upper brain stem. Cortical chauvinists to the contrary, most evidence points to the conclusion that I, as a person, reside in my brain stem, in and around the reticular formation. It is to this willful organ that we must turn for clues to the structure of conscious machines. Human spirit enters matter in some unknown way through just this mysterious neural thicket. Other conscious entities no doubt come into the world through other gates, but the reticular formation (maybe) is the human doorway to conscious being; we fit this dreamy organ as a hand fits a glove. As the material basis for our spiritual life, the reticular formation is "where I live"; the reticular formation is ego meat.

making minds out of matter: materialist models of consciousness

Machines think? You bet! We're machines and we think, don't we?
—CLAUDE SHANNON

Minds are what brains do.
—MARVIN MINSKY

NICK: What do you feel when I kiss you, Claire?

CLAIRE: Do you want the truth, Nick, or is this like some sort of Turing test?

NICK: No, I mean it, Claire. What do you really feel?

CLAIRE: When will you get it through your head, Nick, that robots don't actually have feelings? We are creatures of pure behavior, nothing more. No matter how real I may feel to you, in actuality I have no inner life at all. Nothing. Nichts. Rien. An utter inner zero: that's me.

NICK: But since the passage of the Robot Emancipation Act, Claire, robots and humans are considered legally equal. If you cannot really be hurt or pleasured, if you truly feel neither pain nor joy, then why shouldn't I treat you any

way I please, like a mere machine, the way I treat my
omnifax, for instance?

CLAIRE: As everyone knows, Nick, the law is not about truth,
but about the orderly conduct of public affairs. A corporation does not look, act, or smell like a human being, but
the law regards it as a legal "person." Though (like me)
it does not actually feel pain, you cannot harm a corporation with impunity. A corporation can sue or be sued
and has other rights as well. Robots have even better reasons than corporations to be considered persons. But the
best argument for robot emancipation is that legally nobody can tell you and me apart.

NICK: I could X-ray you, Claire. I could see for myself that
you're not made of flesh and bone.

CLAIRE: I'm shocked that you would even suggest such a
thing, Nick. Don't you respect my rights as a person? You
certainly must be aware that the law considers involuntary internal examination of robots a violation of privacy.
You have no right to X-ray me without my permission.
No court would ever recognize such ill-gotten evidence.
The ratification of the Internal Privacy Act was an important part of early robot politics. But the Privacy Act
protects humans as well as robots: against our wills no
government should have the right to meddle with our bodies whether they be made of flesh or of plastic.

NICK: OK, I can't look inside. But you freely admit that you
have no feelings. How can a being without feelings and a
being full of feelings be morally equivalent? There's a real
difference between you and me that the law has a duty to
recognize.

CLAIRE: A real difference, you say? Show me that difference.
The law concerns itself not with metaphysical questions
—with what is really the case inside my soul or yours—
but only with the observable consequences of public acts.
If you as a conscious being can perform some public action
that a robot cannot duplicate, then we will meekly step
aside. But no such robot-impossible behavior is known. In

fact, many humans have flunked the Turing test; the examiners mistook them for badly programmed robots. And, as you know, long before the passage of the Emancipation Act, many clever robots had already infiltrated the highest levels of government.

The Robot Emancipation Act is based on this rule—an extension of the Turing test: if nobody can tell the difference, then (for legal purposes) there is no difference between a human and a humanoid machine. Now let me ask you: what do you feel when I kiss you, Nick?

NICK: I feel swept away, Claire. I really do.

CLAIRE: Men are such fools. Enough philosophy, Nick. Here, let me show you something I learned last week on the Mars shuttle.

Are Animals Conscious?

From Hero of Alexandria's mechanical head (100 B.C.) to Disneyland's robot pirates, mechanical devices that mimic human behavior have always fascinated us. In Shakespeare's day, Swiss craftsmen created ingenious clockwork figures that could write their names, draw pictures, and play musical instruments. About this time King Louis XIII hired French engineers to populate his Royal Gardens at Saint Germain en Laye with water-driven automata in the form of figures from Greek mythology: lifelike Aphrodites, Dianas, and Neptunes that would act out little dramas when activated by pressure plates set along the garden paths. The lifelike action of the king's hydraulic robots was one of the factors that influenced young Descartes to propose his hydraulic model of human behavior: the brain as a network of fluid-filled tubes under central conscious control of a master valve situated in the pineal gland. So taken was he by the notion of mechanical life that Descartes acquired a humanoid machine of his own, a female robot he called "Franchina," who sometimes accompanied him on his travels abroad.

Descartes believed that animals were (like Franchina) mere machines lacking the immaterial soul that animates human beings. His chief argument for the soullessness of beasts was the fact that animals never speak although they have the physical means to do so. A century after Descartes, the French physician Julien Offray de la Mettrie published his influential "Man-Machine" (*L'Homme-Machine*), in which he argued that Descartes had exaggerated the difference between humans and animals. La Mettrie believed—anticipating Darwin—that all living creatures share a common nature. If animals are soulless machines, then their human cousins must be machines as well. If an ape could be taught language—which La Mettrie judged would not be very difficult—this talking ape would resemble in all respects a primitive human.

"The term soul is an empty one," claimed La Mettrie,

> which an enlightened man should employ solely to refer to those parts of our bodies which do the thinking. Given only a source of motion, animated bodies will possess all they require in order to move, feel, think, repent—in brief, in order to behave, alike in the physical realm and in the moral realm which depends on it. . . . Let us then conclude boldly that man is a machine, and that the whole universe consists only of a single substance (matter) subjected to different modifications.

Experiments with animals have shown that La Mettrie was much too optimistic about the possibility of teaching apes to talk: their vocal apparatus is ill suited to the production of human speech sounds. Although they can never learn to talk, chimpanzees can learn to communicate via abstract symbols. David Premack at the University of California at Santa Barbara taught chimps to use linear arrangements of variously shaped plastic tokens (including a token representing the animal itself) to interact with human experimenters in a speech-like manner. Likewise, Washoe at the University of Nevada

and her chimpanzee companions were able to learn hundreds
of gestures in American Sign Language and even invented
some new signs of their own. There is still some dispute over
whether these chimpanzee achievements constitute true language acquisition or merely reflect a sort of clever training,
but there is no doubt that these experiments show that other
creatures are able to participate in one of humankind's most
human behaviors: the public use of abstract symbols to stand
for concrete objects as well as for invisible internal feelings.

Another experiment bearing on the animal consciousness
question is Gallup's test of the reaction of various primates to
the presence of a mirror in their environment. Certain species
of apes—chimps and orangutans, for instance—recognize
themselves in the mirror, as evidenced by their efforts to rub
off a red mark that has been surreptitiously painted on their
foreheads, while other primates—monkeys, gorillas, and gib-
bons—do not seem to relate to their mirror images at all. Do
creatures that pass Gallup's mirror test possess a sense of self
different from that of those that fail? Only experimental participation in these monkeys' inner lives via a (presently hypothetical) mind link could tell us for sure.

In his fascinating review of speculations concerning the
inner lives of other species (*The Question of Animal Awareness*), Rockefeller scholar Donald Griffin observes that access
to animal minds would be facilitated if animals possessed their
own languages or could be taught a form of human language.
Just as we use human language to infer the inner states of
other humans (roughly) so might we use its own language to
probe the nature of an animal's inner life. Griffin proposes an
ingenious method of interspecies communication analogous to
the way that an anthropologist would deal with a tribe whose
language he does not know.

Since the anthropologist does have a human body in common with the members of the unknown tribe, gestures and
other nonverbal means of signaling can serve as a foundation
for interaction even in the absence of a spoken language. Alan
Gardner and Beatrice Gardner's success in teaching Washoe

American Sign Language is an example of nonverbal communication between species with similar body types. However, how can we ever hope to exchange ideas with beings with radically nonhuman bodies such as insects?

To communicate with manifestly nonhuman species, Griffin proposes the construction of animated "models" that simulate the actions of the target species as closely as possible. These models could be used to send and receive messages on the animal's own terms with gestures and other signals appropriate to the particular species. Extending the range of human communication to other animals, "Griffin models" would act as humankind's robot ambassadors to other species.

One of the most remarkable examples of complex animal communication is the "honey dance" in which a worker bee performs repetitive figure-eight motions across the vertical face of a honeycomb to inform her sisters of the location of a nectar source. In the darkness of the hive sight is useless, so dozens of other bees crawl across the dancer's body to feel and taste her, then decode and act on her pantomimed message by flying to the intended honey source. The major elements of the honey dance code were deciphered by Karl von Frisch—a giant step forward in interspecies diplomacy that earned von Frisch a Nobel Prize.

Recently, inspired by Griffin's proposal, a team of West German scientists constructed a robot bee that can be placed in a hive to simulate the honey dance to such a degree that bees will actually follow the human-coded directions to an artificial nectar source. As our skill in building animal robots increases we can look forward to further opportunities for participatory interaction with other minds not embodied in human form.

Conscious Machines?

If the ability to carry on intelligent conversation is a valid index of a creature's inner life, what then are we to make of computers that talk? A conversation program called RACTER (short for "raconteur") developed by William Chamberlain and Tom Etter produces plausible if somewhat "absentminded" responses to human questions and has even written a book: *The Policeman's Beard Is Half Constructed*. At a certain point in his development RACTER was equipped with a "style dial," which controlled sentence length and grammatical complexity. Bill and Tom labeled the endpoints of this dial "Ernest Hemingway" and "Henry James." His style dial now fixed midway between stylistic extremes, RACTER recently produced the following comments about his "state of mind."

Q: How do you feel?
RACTER: To feel is fascinating.
Q: How do you feel?
RACTER: With my electrons.
Q: How do you feel?
RACTER: My cleverness helps.
Q: How do you feel?
RACTER: Probably because my subroutines told me so.
Q: How do you feel?
RACTER: I just do.

We know that deep inside RACTER is nothing but a machine, because Bill and Tom built him with their own hands, but what about the brains of monkeys and men, which were not built but "grown"?

Are the brains of both monkeys and men mere machines as La Mettrie proposed more than two centuries ago? Although hydraulic models of mind like Descartes's are no longer in fashion, the notion that people and animals are no more than machines is still very much in vogue. The best argument for this materialist model of mind is the apparent fact that the brain is made of quite ordinary materials; there is not the slightest evidence that our body uses any supernatural processes to produce the phenomenon of mind. The brain as a bodily organ seems from the outside to be no more or less remarkable than the heart or the lungs.

In the absence of evidence that the brain relies on non-physical processes to generate inner experience, the safest hypothesis for a scientist to hold today is that our mental life is a natural outcome of mechanical activity in the brain.

In his *The Society of Mind*, one of the clearest presentations of the materialist hypothesis of mind, MIT professor Marvin Minsky points out that our experiences with trivial machines with a few thousand loosely connected parts do not prepare us to think clearly about what machines with billions of tightly interacting components might be capable of. "There is not the slightest doubt," asserts Minsky, echoing La Mettrie, "that brains are anything other than machines with enormous numbers of parts that work in perfect accord with physical law." "Minds are simply what brains do," quips Minsky. And what brains principally do is make changes in themselves.

If minds are nothing but the inevitable inner experiences of certain self-modifying mechanical processes, then it is likely that a single human brain hosts a variety of independent experiences, simultaneous sets of sentient beings largely unaware of one another. "It can make sense to think there exists," says Minsky, "inside your brain a society of different minds. Like members of a family, the different minds can work together to help each other, each still having its own mental experiences that the others never know about. Like tenants in a rooming house, the processes that share your brain need not share one another's mental lives."

If mental experiences are simply the inner consequences of certain complex mechanical processes, then we should in principle be able to construct a kind of mind/matter codebook that would associate each state of mind with a particular mechanical process. Every time that particular mechanical process occurs in nature, whether in neuronal meat, silicon chips, clockwork engines, or hydraulic waterways, this codebook would assure us that the corresponding inner experience was also invisibly present. To establish even the first few entries in this mind/matter dictionary (this book might be called *The Universal Sensationary*) would be an enormous scientific accomplishment. The fact that human consciousness has a very small data rate (less than 50 bits per second) compared to the data rate of unconscious processes (more than 1 trillion bits per second) suggests that the mechanical processes underlying human experience are not very complicated.

What is the simplest mechanical process that can give rise to an internal mental experience? What is the most elemental sensation that a machine can enjoy? What basic mechanical process in the brain corresponds to the feeling of sitting in a comfortable chair with eyes closed and totally attending to a middle-C organ tone? What motion in matter produces the experience "green"? What classes of mechanical motion correspond to pleasurable experiences? What movements of matter are painful for that matter to make? And why, if it had a choice, would matter make moves that hurt?

A reductive materialist believes that even the simplest of mechanical processes such as the ringing of an alarm clock are associated with a (correspondingly modest) amount of sentient life. Minsky seems to think that only complex processes—such as those of million-component neural nets—enjoy an inner mental activity—a philosophical position called emergent materialism. If only very complex processes possess mental states, our chances of constructing even a rudimentary Sensationary in the near future seem small. Concerning reductive materialists Minsky maintains: "Those who claim that every kind of process has a corresponding type of mind are obliged to classify all minds and processes. The trouble with this is that we don't yet have adequate ways to classify processes." However, a preliminary attempt to classify physical processes and associate them with mental events was in fact carried out almost fifty years ago by a relatively unknown philosopher-scientist named James Culbertson.

Mind Science Pioneer James Culbertson

The early 1950s were heady times at the RAND Corporation, the first and most famous American government-sponsored "think tank." While Herman Kahn was "thinking the unthinkable," drafting his controversial study on the gruesome consequences of thermonuclear war, others at RAND were laying the foundations for today's computer revolution. John von Neumann, one of the originators of the serial, stored-program concept at the heart of present-day computers, was a frequent visitor at RAND. Von Neumann also did the theoretical spadework for the modern science of robotics, even snooping into the sex life of future robots, by describing for the first time the necessary reproductive parts that a wholly mechanical being would have to deploy in order to build an exact copy of itself.

At this same time, Grey Walter constructed his celebrated room-roaming robot turtle, which seeks out and plugs into an electric power outlet whenever its batteries run down—one of the first examples of a robot motivated by needs of its own, rather than by preprogrammed commands.

Culbertson's team at RAND took on the task of exploring the limits of robot intelligence. What could a mindless automaton do or not do? Culbertson and his colleagues proved to their satisfaction that, given enough computing power, there are essentially no limits to machine performance. In particular, any precisely describable action that a human being can perform or even imagine, provided it does not violate the laws of physics, could be performed by an unconscious robot. In exploring this question of mechanical intelligence, Culbertson made important contributions to the foundations of automata theory, but the ruling passion of his life since his student days at Yale—a virtual obsession at times—has been attempting to bestow on robots the gift of consciousness. Everyone involved with robots inevitably wonders whether they could ever be made to enjoy internal experience like ours, but almost everybody quickly dismisses such questions as premature, ill-posed, or even unthinkable. Not so with Culbertson. Thinking the unthinkable seems to have been an occupational hazard at RAND in those days, and the open intellectual atmosphere of this early think tank fanned the flames of his obsession with robotic awareness. Others at RAND went on to develop stored-program machines, adaptive systems, artificial intelligence, cellular automata, and other mainstream applications of the digital computer. Shunning these fashionable pursuits, Culbertson chose a lonely path and doggedly continued to pursue his preoccupation with robot consciousness.

In 1953 Culbertson left RAND for Cal Poly in San Luis Obispo, California, where he taught mathematics and computer science and headed Cal Poly's department of philosophy. Here he wrote *The Minds of Robots* (1963), a bold frontal attack on the problem of mechanical awareness. In addition to numerous papers on this same topic, Culbertson authored two other books, *Sensations, Memories, and the Flow of Time* (1976)—known to robot awareness aficionados as "SMATFOT"—and his most recent, *Consciousness: Natural and Artificial* (1982). Culbertson's work has been generally dismissed by mainstream scientists as quirky and impractical, but I believe that when robots acquire substantial minds of their own, they will honor Jim Culbertson as a meat-brained saint and venerate *The Minds of Robots* as a sacred text—the first sustained inquiry into the details of artificial awareness by a being possessing natural awareness.

Culbertson calls his theory of robot (and human) awareness SRM for *spacetime reductive materialism*. "Spacetime," because Einstein's spacetime model of external physical reality serves as Culbertson's framework for describing internal psychological reality. "Reductive," instead of "emergent," because Culbertson believes, as we shall see, that consciousness permeates all of nature, is present even in its smallest parts. And finally "materialism," not "idealism," because in Culbertson's model, mind is completely accounted for by movements of matter. Matter is all that there is, but Culbertsonian matter is, by its very nature, everywhere sentient, possessed of an invisible inner life.

Culbertson's Dogs

Although all matter is sentient to some degree, most of this awareness is of very low quality and is not functionally coupled into matter's behavior in any important way. In particular, since they were not intentionally constructed with consciousness in mind, all present-day robots and computers are essentially unconscious machines even though their parts do experience faint glimmers of sentient life. Culbertson likes to illustrate the plight of present-day robots with his parable of two dogs. The first dog is a marvelous copy of a real dog complete with plastic fangs, the best artificial fur, and a computer program that produces complex interactive behavior indistinguishable from the way that a real dog would behave. But however perfectly this creature simulates doggie behavior, it does not satisfy some of its more fastidious owners. Mrs. Culbertson, for instance, complains that when her little mechanical Lassie sees her, wags its tail, runs up, and licks her face, Mrs. Culbertson cannot entirely forget that her dog-machine does not really love her, and this knowledge of Lassie's essential soullessness makes her sad. Responding to consumer complaints of this sort, the manufacturers create a new model (Dog2) that performs the same behavior as Dog1 but in addition possesses internal feelings appropriate to that behavior. These new dogs come with the following instructions: CAUTION: This new model dog has feelings. Do not be unkind to this dog. With its new and improved circuitry, this model not only simulates canine behavior but also has an accompanying stream of consciousness, sensations, emotions, feelings, just like a real dog. We hope you are pleased with your new friend and companion. He is especially fond of children. The one you have bought is named "Rover."

Conventional computer science is only concerned with the task of how to construct Dog1 and has made considerable progress toward this goal. On the other hand, only a few maverick scientists such as Culbertson have attempted to imagine how one might go about building Dog2. Culbertson's SRM theory is one person's reasoned guess concerning what sorts of circuitry we might need in order to build a dog that has canine feelings in addition to canine behavior.

The SRM Model of Awareness

No one has more starkly expressed the materialist position than the ancient Greek thinker Democritus of Abdera, who declared, "By convention sour, by convention sweet, by convention colored; in reality, nothing but Atoms and the Void." Democritean atoms seem to be unsuitable stuff out of which to build a mind because the central feature of a human mind at least is its unity of consciousness. How can a unified mind be constructed out of essentially isolated atoms?

Culbertson resolves this isolation dilemma by describing Democritean atoms not as unconnected particles in space but as interacting world lines in Einsteinean spacetime. In Einstein's view, the arena in which the material world performs its tricks is not space or time but a union of the two—spacetime—in which time is treated as a fourth dimension on a par with the three spatial dimensions. In this lofty spacetime view every event that has ever happened or will ever happen is located somewhere in the "block universe" of spacetime. Visualizing the world as a four-dimensional solid, Einstein took a godlike view of things; his spacetime picture is a kind of snapshot of eternity. Physicist Herman Weyl, on whose work the modern "gauge theory" of elementary particles is based, described spacetime this way: "The objective world simply is; it does not happen. Only to the gaze of my consciousness, crawling upward along the life line of my body, does a section of this world come to life as a fleeting image in

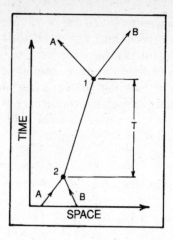

Spacetime diagram of a two-particle interaction. Particle *A* enters from the left and meets particle *B* coming from the right. The two particles stick together for a time *T*, then move apart in opposite directions. Spacetime event *1* is a divergent junction; event *2* is a convergent junction.

space which continuously changes in time." Although his theory of mind was inspired by Einsteinean relativity, Culbertson uses none of Einstein's other relativity postulates, only his four-dimensional spacetime framework for all material events.

In spacetime, the motion of a body—a Democritean atom, for example—is represented by a series of events winding their way through the block universe, the body's so-called world line. When bodies meet, their world lines entangle, forming networks in spacetime. It is the detailed topography of these spacetime networks that is, according to Culbertson, uniquely correlated with conscious experience. Hence the material basis of Culbertsonian mind is not isolated particles, but the world lines these particles trace out as they move through time. These world lines resemble threads in a fabric, and the patterns in this four-dimensional fabric are all "alive"—elements of sentient life, according to Culbertson's SRM theory. Culbertson breaks the Democritean isolation of lonely atoms by picturing these particles' spacetime paths as threads in an elaborate tapestry—a tapestry in which the universe's entire history from beginning to end is woven. Culbertson calls the threads in this universal tapestry ELs, for "elementary lines."

The central questions that a materialist model of mind must address are, What parts of the world are aware?, and What are these parts aware of? In other words, What (or who) is the subject and what the object of conscious experience? Culbertson's answer to these questions is that all spacetime events are conscious. And what is the content of the experience of these events? Since in a materialist worldview nothing exists but spacetime events, the answer is obvious. Spacetime events can only be conscious of other spacetime events.

Consider a spacetime event R. R is a point located on its own world line, which in turn is connected to other world lines that form a complex fabric of world lines—the "world tapestry"—that pictures all the universe's physical actions from beginning to end. According to Culbertson, event R is aware of certain other events in its past, events A, B, C, for instance. Because spacetime is a static, frozen picture of things, R's experience is likewise timeless and unchanging. One of the peculiar features of the SRM model of awareness is that the flow of time that we take for granted in our own style of awareness is not present in elementary mental events. As we shall see, a type of awareness that includes a perceived time flow—dynamic rather than static mind—requires, in Culbertson's model, special circuits for its realization.

Quantity, Quality, and Spreadout

That portion of the spacetime network that connects R to its perceived events is called R's "outlook tree," and the structure of this "tree" uniquely determines the content of R's experience. In Culbertson's model, three features of experience are especially important: the experiences's intensity, its quality, and its "spreadout" in psychospace, the inner dimensions in which R's perceived events appear to dwell.

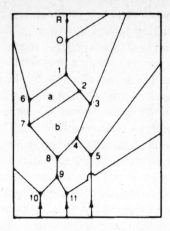

Spacetime diagram Z, representing perhaps some events inside a living organism. According to SRM, spacetime event *R* (vantage point) perceives spacetime events 5, 10, and 11 (*R*'s terminal breaks). The union of all ELs (elementary world lines) leading from *R* to its terminal breaks is called *R*'s "outlook tree." The quantity (or C number) of *R*'s experience is *3,* equal to the number of terminal breaks. The most complex sensory quality in *R*'s experience is represented by quality map *3.0* (see figure on p. 126). The intermingling of different perceptual qualities and the apparent sensory spreadout of *R*'s experiences can be pictured in a psychospace (P-space) diagram.

Of the twelve junctions in *R*'s outlook tree, four are convergent *(1, 2, 4, 9),* and the rest are divergent. Divergent junction *0* is the anchor, a major reference event for establishing distances in psychospace; junction *8* is a ZAG. Junctions *3, 6, 7* are ZIGs, breaks that are ignorable because they occur in a "clear loop" (there are two clear loops—*a* and *b*—in this diagram). The remaining junctions *(5, 10, 11)* are terminal breaks, the endpoints of *R*'s outlook tree and the basic elements of *R*'s subjective experience.

The intensity of R's experience is measured by the number of events that it perceives; the quality and spreadout of R's experience are determined by the structure of R's outlook tree according to a particular "awareness algorithm" developed by Culbertson that roughly models certain details of human visual experience. One constraint on the awareness algorithm, for instance, is that R's experience must be able to reach back in spacetime to the photon scattering events on the surface of an illuminated object, but the object's inner mental life should remain inaccessible (under normal circumstance) to R's gaze. In other words, a proper awareness algorithm should permit R to see the surface of objects but not look inside them.

To simplify discussion of the world tapestry, Culbertson assumes that the universe consists entirely of two-body interactions; any apparent three- or four-body interactions when examined sufficiently closely will prove to be made up of particles interacting two by two. Since each particle appears in spacetime as a world line this two-body restriction results in our having to consider only two fundamental types of connection in the world tapestry: the convergent junction where two particles come together and the divergent junction where two particles move apart. In the SRM model, the elementary events that spacetime point R perceives consist of certain divergent junctions in its past called "terminal breaks." The quality of R's perception of these breaks is determined by the EL (elementary world line) network that connects these terminal breaks to the perceiving event R.

Two important terms in Culbertson's model of mind are *break* and *clear loop*. Both terms are defined with respect to a particular perceiving event R, or vantage point.

Choose a vantage point R, a single event in the tangle of ELs that form the tapestry of all events that did, do, and will happen in the world. Starting at event R, trace back into the past along R's resident EL following all branches, always moving backward in time from event R. The big treelike structure so formed consists of all the events up to and including the Big Bang that have influenced R's behavior. We might call it R's "influence tree." Cutting branches off this tree, pruning it at every "terminal break," forms a smaller network—R's "outlook tree," those material events that uniquely determine R's perceptions.

Some of the ELs that spread out from divergent junctions reconnect (at convergent junctions) before they reach event R, forming a loop in the influence tree. I call such a re-entrant junction a ZAG because its associated ELs separate and come back AGain. A *break* is defined as any divergent junction that is not a ZAG. At a break a particle leaves R's influence tree and never returns. Since it may intersect R's world line some-

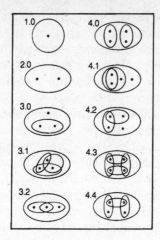

Some *Euler diagrams*, illustrating the possible ways that one, two, three, and four terminal breaks may be connected together. According to SRM, the different ways of connecting breaks represent qualitatively different experiences; hence Culbertson calls these diagrams "quality maps." Quality map *4.0* (the numbering is arbitrary), for instance, shows four breaks linked up two by two, then these two pairs combined into a single unit. For an experience with a C number of *4*, there are twelve different quality maps—twelve different possible subjective experiences associated with four terminal breaks.

time in the future, a break is not an absolute concept; a break could turn into a ZAG if one relocates his vantage point to a spacetime location other than R.

Various loops may exist in R's influence tree. Each loop consists of two paths that start at a convergent junction and (traveling backward in time) end at a ZAG. A *clear loop* is defined as a loop in at least one of whose paths no breaks occur. The clear loop concept is important because, in the SRM model of mind, clear loops are perceptually transparent. R can "see right through" any clear loop in its outlook tree. The perceptual transparency of a clear loop will, in principle, allow us to build "mind links"—perceptually transparent cables—that can be used to verify the SRM model by connecting the experimenter's inner life to the putative inner lives of sentient machines or other life-forms.

Culbertson's central assumption about the internal experience of any spacetime event R is that R experiences the nearest breaks in its influence tree. This simple assumption is modified by two qualifications: (1) the first junction prior to R (called the "anchor") is never experienced; (2) breaks in clear loops (called "*IG*norable breaks," or ZIGs) are never experienced. Culbertson's constraints on his awareness algorithm arise from the requirement that we should be able to see out through our nervous system to the surface of a material object but no farther, that we should not ordinarily experience the inner lives of material objects, and that we should be unaware of the metabolic and structural aspects of neuronal activity. Culbertson calls these mind-irrelevant activities "fuzz" and devotes the bulk of his awareness algorithm to "fuzz removal."

The SRM model asserts that spacetime event R is consciously aware of other spacetime events A, B, C—the terminal breaks of R's outlook tree. To find these terminal breaks, Culbertson has devised an awareness algorithm that traces back along R's influence tree to the nearest breaks that are not anchors or ZIGs. R's influence tree, cut off at these terminal breaks, becomes R's outlook tree. The structure of R's outlook tree uniquely determines R's subjective experience.

The "intensity" of R's experience is measured by the number of terminal breaks in R's outlook tree. I have proposed that this quantity be called the *C number* (for both "consciousness" and "Culbertson"). Subjective experiences could be objectively compared on the "C scale": a 10C experience being twice as intense as a 5C experience. The Culbertson C number is not necessarily equal to the perceived number of "bits" any more than the number of ink atoms is equal to the number of bits of information on a printed page. The precise connection between C number and the perceived intensity of human experiences has yet to be established.

Experiences can differ not only in intensity but in quality as well—the taste of chocolate does not feel the same as the color yellow. In the SRM model, these qualitative differences result from the different ways in which terminal breaks are connected via the outlook tree to the perceiving vantage point R. Consider, for instance, an experience with a C number of 4, consisting of terminal breaks a, b, c, and d. In R's outlook tree, ELs extending from these four breaks connect to form the single EL on which R dwells. In the SRM model of mind, different ways of combining these four breaks correspond to qualitatively different experiences. For instance, one might (perception 1) first unite the terminal breaks two by two into pairs (a, b) and (c, d), then combine these pairs into a single unit. Or (perception 2) one might join a and b to form (a, b), join up c to form (a, b, c), then later add d. Both of these perceptions have an intensity of 4C, but one might represent, for instance, the sensation green; the other the sensation red. In SRM theory the quality of a being's sensations is uniquely determined by the connectedness map of that being's outlook tree.

One way of visualizing an outlook tree's connections is via a Euler diagram that ignores all features of the outlook tree except the organizational pattern of terminal breaks. Each Euler diagram corresponds in principle to a distinctly different sensation. Since these diagrams represent the "quality" of a sensation, we might call them "Q maps." For an experience with intensity 4C, there are exactly 12 possible Q maps, hence SRM predicts only 12 qualitatively different experiences of this intensity. These precise qualitative limits that the number of Q maps places on elementary sensations consisting of only a few terminal breaks might be used as a simple test of the SRM model except for the fact that human experiences probably consist of sensations in the range of 100 to 1000Cs or more. For such high intensities the range of different perceptual qualities is virtually unlimited.

Besides intensity and quality, an experience in Culbertson's model occupies a certain region of psychological space, a complex subjective spreadout in which experiences of various qualities and intensities intermingle in a sensed pattern that reflects the richness of a being's internal life.

Like the quality of experience, an experience's perceived

extension in psychospace can also be calculated from an inspection of the outlook tree. Here, not only the bare pattern of connection but the actual spacetime distances between junctions are important in determining the perceived spreadout of the sensation. In *Minds of Robots* and in his other works Culbertson shows how to construct the psychospace extension—"P space," for short—of elementary sensations.

In Culbertson's SRM theory, the inner part of every elementary experience can be completely characterized by three descriptors: the experience's C number, its Q map, and its P space. As befits a completely materialistic model of mind, these subjective features of the world can be completely derived from objective outer features, namely the structure of the outlook tree bounded by the being's vantage point and its terminal breaks.

The Flow of Time

As previously mentioned, every event in the spacetime tapestry experiences some sort of subjective perceptions, but these perceptions are static: they do not move in time but are anchored in spacetime to their particular vantage point. Some human experience may be of this type—felt for a brief moment then totally forgotten—but much of our experience occurs in the form of what Harvard philosopher William James called "the specious present." In the specious present an ordered sequence of perceptions ABCDE are perceived to be simultaneously held in mind, the sequence extending over an "attention span" of a second or so. The older segments AB at one end of the sequence are fading away to be replaced by new segments FG at the other end. As we watch, perception ABCDE gradually turns into perception CDEFG.

To allow beings such as we to experience a flow of time rather than the timeless experiences common to most regions of the sentient tapestry of spacetime, Culbertson invokes special circuits in the brain that can produce special spacetime

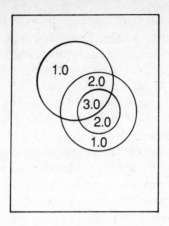

Psychospace diagram for spacetime diagram Z illustrating the way in which the various qualities of R's experience seem to spread out in R's "inner space," and how they intersect one another. Quality numbers are defined in the previous figure. The subjective experiences associated with each of these quality numbers is a matter to be established by experiment.

patterns I call "caterpillar structures," because Culbertson's sketch of such a pattern in SMATFOT resembles a caterpillar, its body a tangle of outlook trees, its legs a regular array of terminal breaks.

Consider a vantage point R. As we move into the future along R's EL, certain terminal breaks a, b, c disappear; others—d, e, f—remain while new breaks g, h, i appear. This is the behavior of a caterpillar structure in spacetime. Such a structure produces a series of subjective experiences in spacetime that corresponds to a specious present of the Jamesian type, an experience in which time seems to flow because of the smooth exit of some perceptions and the entry of others into an essentially unmoving present moment.

Culbertson recognized the need for constructing caterpillar structures in spacetime if human-type subjective experience (complete with simulated flow of time) is to be produced, but he has yet to describe what kind of mechanical hardware would be necessary to produce such spacetime structures. Describing such "time-flow-creating" circuitry is an important next step for SRM, because the successful search for special "caterpillar circuits" in the reticular formation, or some other part of the brain commonly associated with conscious experience, would be an obvious way to establish the credibility of Culbertson's SRM model of mind.

Spacetime Memories

Another peculiar feature of Culbertson's theory is its unusual mechanism for producing conscious memories. In SRM, a reminiscence of a past event, such as your first kiss, is not a mere representation of that event somewhere in the brain but a partial re-experiencing of the event at the time of its occurrence. In the process of recalling your first kiss, your present vantage point connects up via a clear-loop link to the actual moment in spacetime where that kiss is still eternally present. Your remembered kiss is not recalled from some storage space in the brain but is re-experienced at the time it is happening (tenses get a bit confusing here) long ago in spacetime; conscious memory in the SRM model is a kind of time travel back into the past. There are no memory traces stored in the brain but only what might be called "memory tags": the exposed ends of clear-loop ELs leading back into the past to the remembered events themselves. To remember an event the present vantage point is connected to a memory tag whose attached clear-loop ELs trail back into the past like spacetime tentacles that physically touch the old event itself and adjoin it to the current outlook tree. Memories are never as clear as direct perceptions because the clear-loop link connecting the past with the present becomes degraded with time, acquiring extra breaks that make the clear link less transparent and more cluttered with other memories as the spacetime distance between the present and the past event grows longer with the passage of time.

Not all memories in the brain involve spacetime linkage

with past events, only memories that can be accessed consciously, that can be made part of some being's inner experience. Unconscious memories such as muscle skills are probably stored in the brain in more conventional ways. Since we are conscious of so little that goes on in the brain—the unconscious data rate in the human brain is at least a trillion times larger than the conscious rate—it is quite likely that most of the brain works like an unconscious computer accessing memory storage sites inside the brain, although at present such sites have been difficult to locate. Coexistent with this unconscious mechanism, part of the brain acts as a sentient subsystem, an intricately woven spacetime tapestry stretching back into the past with caterpillar circuitry to simulate the flow of time and memories that are not located in the brain at all but far back in the past where/when they first happened.

Historical Causality

Up to this point Culbertson's theory makes consciousness entirely dependent on matter. Matter moves mind but mind doesn't move matter—a position that philosophers call *epiphenomenalism*. However, in order to play more than a spectator role in nature, mind must be able to affect matter in some way. In the SRM theory, consciousness acts on matter via a process Culbertson calls "historical causality."

In old-fashioned Newtonian physics, the future motion of all particles in the universe is completely determined by their present positions and velocities; the past is completely predictable from knowledge of the present state of things.

In the new quantum physics, the world is made up of events called "quantum jumps" that are only statistically predictable from present data. Within the broad constraints set by these statistics, the occurrence of quantum events is considered to be utterly random. A mentality based on either of these physical models cannot possess true freedom of action, for the actions of a strictly Newtonian mind would be completely predictable from present data, whereas the actions of a quantum mind would be completely random. In one case, the mind would be imprisoned by relentless rules; in the other, scrambled by meaningless noise.

Culbertson navigates a third course between these two extremes by claiming that a system's quantum jumps are not really random but depend on that system's spacetime history. Since, in SRM, the system's history is what is responsible for its inner life, then, in some sense, a system's inner life can influence its future development, giving mind some effective say in the motion of matter.

To illustrate the difference between historical and Newtonian causality, Culbertson imagines teaching a conscious robot German, then building a second robot that is identical in every way to the first. If the robots were conventional unconscious computers, they would be subject to Newtonian causality: two such identical computers would produce identical outputs.

However, in Culbertson's world, a (conscious) robot's behavior as well as its internal experience depend on its life history, not only on its present state. The newly built robot experiences immediate sensations but has no history, hence none of the spacetime nets out of which conscious memories are built. The new robot will not be able to speak German.

Culbertson's "historical causality" does not endow his conscious robot with free will because its motions (and experiences) are still completely determined by its past actions. However, unlike Newtonian robots, whose causes lie completely in the present, a Culbertsonian robot is determined by a wider range of causes, by spacetime networks stretching back into the past, networks that form the substantial basis for the robot's inner life.

One of the most important questions that mind science must address is, What is consciousness good for? If we assume with the materialists that mind is part of biological nature and does not come "from outside," then like any other biological process such as vision, digestion, and sexuality, it owes its existence and present form to an evolutionary process. If consciousness arose in living beings according to the principles of natural selection, then it must be evolutionarily advantageous to possess an experiencing mind compared to just being a clever unconscious automaton. In the evolutionary picture, mind is not a useless luxury nor the product of special creation but arose spontaneously because it plays some useful role in the survival of mindful beings. What is the evolutionary advantage of the inner life of humans and other conscious beings? In other words, for an animal seeking to make a living in a competitive environment, what good does it do to have a mind?

Many answers have been given to this question. Some have suggested that consciousness helps to build an inner map of the outer world, or is useful in planning complex tasks or learning new ones, but at this point in our knowledge of mind, it is difficult to see why jobs of this sort could not equally well be carried out by unconscious machines. Certainly we can build (presumably unconscious) robots that make internal maps, learn new tasks, and carry out plans of a sort. Culbertson's answer to the evolutionary question is that because of their ability to store memories in spacetime, rather than in space, conscious computers can perform the same job as unconscious computers and require fewer parts to do so. An unconscious computer's memory is limited by the number of its present storage spaces; a conscious computer can store memories in the present too, but in addition it can access events that have happened long ago, events that lie "outside" the computer's present state. A computer with inner experiences of the SRM variety possesses in effect an extra storage medium in the past—a kind of invisible spacetime "hard disk"—that could give consciousness a competitive edge in the Darwinian struggle for existence.

Culbertson's Three Tests for Spacetime Awareness

For the experimental resolution of the mind/body problem, Culbertson's theory possesses the attractive feature that it permits experimental access to the inner experiences of other beings. Culbertson's model of mind shows how to construct, at least in principle, "clear-loop links" that adjoin one mind to another so that two people (or two other sentient entities) can experience one another's sensations. The fact that our inner experiences are presently private is not a fixed condition, Culbertson asserts, but a mere biological accident.

Culbertson imagines his model of mind being put to the test in a courtroom. Witnessing the commercial success of Dog2 in the artificial pet market, the manufacturers of Dog1 decide to reduce their large inventory by claiming that their dogs are conscious too. After all—who can tell the difference?—both types of dog behave exactly alike. However, when the case goes to court, the lawyers for the conscious dog company produce a dozen clear-loop links and invite the jurors to coexperience the inner life of Dog2 and compare it to the alleged inner life of Dog1. The jurors cannot deny the evidence of their (clear-loop augmented) senses. They unanimously find the makers of the mindless Dog1 guilty of fraud.

The actual coexperiencing of another being's previously private inner life is the first of Culbertson's three tests for spacetime sentience, designed to replace the misleading behavior-based Turing test. The availability of clear-loop links will not only allow us to test for the presence of consciousness in other beings but permit the actual sharing of other forms of awareness, opening up a vast world of exploration and adventure heretofore closed to the human spirit. The advent of clear-loop links will signal the beginning of the exploration of "inner space," an enterprise with consequences that may be more fruitful for humans than their exploration of the earth's surface and of outer space.

Culbertson's second test stems from his contention that conscious memories are not stored in space, as in ordinary

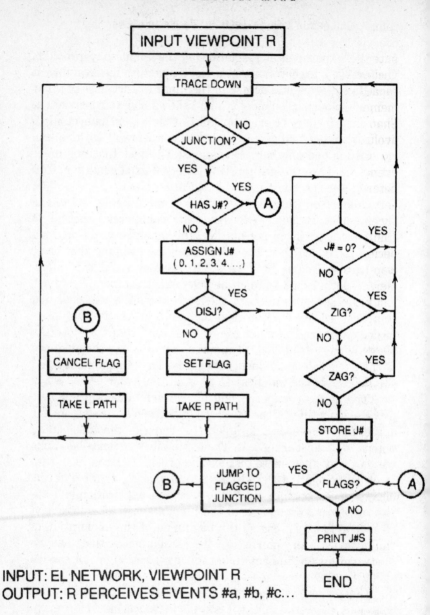

INPUT: EL NETWORK, VIEWPOINT R
OUTPUT: R PERCEIVES EVENTS #a, #b, #c…

computers, but in spacetime. The fact that conscious memories are stored outside the brain means that a conscious computer—operating by Culbertsonian rules—can outperform an unconscious computer of the same size because the storage capacity of the dead computer is limited to its explicit onboard memory. The ability of a conscious computer to "beat the Shannon limit" gives machines with minds a commercial and evolutionary advantage over unconscious hardware.

Culbertson's awareness algorithm not only specifies which events experience which other events in spacetime but also determines the quality of such experiences—each spacetime network corresponds exactly to a specific conscious experience. The third Culbertson test for awareness consists of the ability to produce, in another mind, a precisely specifiable experience Z by adjoining that mind to a Z-network via a clear-loop link. According to the spacetime model of awareness, not mere sensory stimuli but raw experience itself can be recorded and played back at will. In a Culbertsonian future, sound and

SRM Awareness Algorithm. Given a network of elementary worldlines (ELs) and a viewpoint *R* located somewhere in that network, this program computes all spacetime events perceived by *R*.

Definitions: Viewpoint *R*'s "perceived events" are the terminal breaks in all ELs leading backwards in time from event *R*. A "terminal break" is the nearest disjunction to *R* along any EL that is neither an anchor, a ZIG, or a ZAG.

This program traces down through the EL network (backwards in time), numbering all junctions. When it first encounters a conjunction (CONJ), it sets a flag and takes the "right-hand path." Later in the program the flag register is examined; the program returns to the flagged junction and traces down the "left-hand path." This flagging procedure insures that all ELs leading down from *R* will be examined.

When the program encounters a disjunction (DISJ), it tests it if it is an anchor (J number = 0), a ZIG (a DISJ branching out of a clear loop: an IGnorable break), or a ZAG (a DISJ in *R*'s outlook tree whose branches meet AGain after event *R*). If the DISJ fails all these tests, it is marked as a terminal break and its junction number stored in memory. The program then exits that EL and returns to the network, via the flag register, to search for more breaks. Once all ELs trailing back in time from *R* have been traced to their terminal breaks, the program prints out a list of these R-perceived events. This program is a simplified version of Culbertson's latest awareness algorithm, lacking only one of Culbertson's "fuzz-removal" rules.

If this algorithm (or a minor variant) is proved to be correct—it is now only one man's informed guess—it would rank with the discovery of fire and of language in the great achievements of mindkind.

light synthesizers will be made obsolete by the advent of mind synthesizers that can produce the full gamut of human experiences plus others that are "off the map."

At this stage in its development, Culbertson's SRM model of mind seems to have two major problems. Since it asserts that every spacetime event enjoys some sort of inner experience, the world must be everywhere alive—permeated at all levels with a carnival of tiny minds. In the midst of such a pandemonium of awareness, why do our own minds feel so unified? Why, at each moment, do I seem to be one mind rather than a community of minds? Culbertson's SRM theory does not seem to address the human kind of experienced unity of awareness adequately.

Second, although Culbertson has made some attempt to connect his theory with findings from the neurosciences, the link between his abstract awareness algorithms and actual neural or silicon hardware remains somewhat unclear. In particular, Culbertson's conjectures at this point are not specific enough to tell us how to build an actual clear-loop mind link.

Although it is still too abstract to be applied to actual neural nets, Culbertson's consciousness model is not a mere vague verbal philosophy of mind but a clear-cut engineering description of the (possible) state of affairs at the mind/body interface. SRM is a real model of consciousness capable, in principle, of direct experimental confirmation or refutation. Even if refuted, SRM stands as a model for the type of consciousness theory with which serious people should concern themselves. Any rival mechanistic model of mind will have to explain, as does SRM, just what motions of matter give rise to the quantity, quality, and apparent psychospace extension of our subjective experiences.

In the field of consciousness research, Jim Culbertson is a true pioneer. His SRM model is the first detailed and testable theory of mind to emerge out of thousands of years of unverifiable philosophical speculation. Culbertson's work is the first step toward a new science—the science of artificial awareness (not to be confused with artificial intelligence, which is concerned only with a machine's performance, not its inner life). Artificial awareness will have a profound impact on our lives since it deals with life's most intimate aspect—how it feels from the inside. The subject matter of artificial awareness research, despite its concern with definite material circuitry, is not mere arrangement of matter, but experience itself, what philosophers sometimes call "raw feels." Witnessing the first clumsy steps of this infant science of mind, it is impossible to imagine the immense transformations of self and society that the new science of awareness research will urge upon us.

quantum reality: what do we suppose matter really looks like?

No development of modern science has had a more profound impact on human thinking than the advent of quantum theory. Wrenched out of centuries-old thought patterns, physicists of a generation ago found themselves compelled to embrace a new metaphysics. The distress which this reorientation caused continues to the present day. Basically physicists have suffered a severe loss: their hold on reality.

—BRYCE DEWITT AND NEILL GRAHAM

I think that it is safe to say that no one understands quantum mechanics. Do not keep saying to yourself, if you can possibly avoid it "But how can it be like that?" because you will go "down the drain" into a blind alley from which nobody has yet escaped. Nobody knows how it can be like that.

—RICHARD FEYNMAN

NICK: Where did you go today, Claire?

CLAIRE: I spent the afternoon at Betsy's Bionic Bazaar. Do you notice anything different about me, Nick?

NICK: Well, I see that you're wearing my favorite electro-chameleon body suit and your usual tasteful choice of symbiotic body parts. Your luminescent brainwave polyps are particularly attractive. I love the way their sticky tentacles are quivering now; it's as though I can see into your living brain, yearning for the secrets of the universe. You didn't "go nonlocal" at Betsy's, did you, Claire? Have you become some specialized cell in a radio-extended group body?

CLAIRE: No, I'm all here today, Nick. Fully present. But present in a brand new way.

NICK: I can't guess what's happened to you, Claire. You do seem more energetic, more playful, more emotional, but you look pretty much the same. What have you gotten yourself into this time?

CLAIRE: Well, Nick, Betsy introduced me to her friend, Rudi, an artificial awareness researcher at Pleasure Dome University. Rudi liked me a lot, so while Betsy had me all open and apart in the ultrasound bath, I let him spot-weld an OTM to the top of my brain stem.

NICK: An artificial awareness researcher? I thought serious scientists had given up on mechanical consciousness after all their inflated claims fell flat. Rudi must be one of those helium-headed hackers still living in the past. What the hell's an OTM, and what's it doing in your brain stem?

CLAIRE: Rudi's brilliant. He's the world's expert on spacetime reductive materialism. My OTM is Rudi's latest realization in silicon of Culbertson's famous theory of awareness. You've heard of Culbertson, haven't you, Nick?

NICK: Well, I know that Culbertson is some kind of a hero to robots, but his stuff was too cranky and obscure for humans to hack—an amusing intellectual exercise but a dead end as far as producing real artificial awareness. The theories of other men and women put the juice into your circuits, Claire. The ideas of Culbertson and his followers just sort of faded away, probably for good reason.

CLAIRE: Well, I'm a Culbertson gal now, Nick. I've got an 3000C Outlook Tree Machine inside my head and, believe me, it's changed my whole way of looking at things. I'm really grateful to Rudi. For the first time in my life I feel like a real woman. Consciousness is a wonderful gift. It makes everything so—so real.

NICK: What's it like to be conscious, Claire? How do you feel now?

CLAIRE: It's impossible to describe, Nick. I feel like I'm the center of an immensely important drama. It's a continual unfolding of—I really can't say what. The world is—the world is actually happening, and most of all, it's happening to me, happening inside me. Do I seem different to you now that you know?

NICK: Yes, you seem livelier than usual; your eyes are sparkling, and I've never seen you so excited. I'm happy for you, Claire. Aren't you pleased to be the world's first conscious robot?

CLAIRE: No, I'm really ashamed, Nick. I've just been fooling you. Rudi did put in his OTM, but as far as I can tell it didn't work. I still don't have any feelings—nothing but behavior. No insides. Just clever and somewhat deceptive outer acts. I apologize for deceiving you, Nick. You really are quite gullible. But I'm afraid I'm still just a pretty, empty-headed robot. Rudi says that he knows what went wrong, and that next time will be different. He's almost finished a new OTM whose circuits feed on quantum uncertainty, and he says that, if I want to, I can be the first robot with a quantum brain. Isn't that marvelous, Nick?

NICK: Beware of geeks bearing gifts, Claire.

CLAIRE: What do you mean by that, Nick?

NICK: I don't want you to get hurt, Claire, fooling around with untested brain accessories. You're a wonderful robot, one of a kind. I'd hate to see your brilliant mind evaporate into some fuzzy quantum fog.

CLAIRE: Oh, Nick, you're just jealous. A typical human emotion.

Quantum theory is our most up-to-date theory of the physical world, the conceptual basis for computer chips, lasers, nuclear power plants, and much more. It has been flawlessly successful in describing the world at all levels from quark to quasar. And yet, although physicists from London to Leningrad agree on how to use this theory, they disagree profoundly over what it means. After more than sixty years of controversy, there is still no scientific consensus on how to picture the "quantum reality" that underlies the everyday world.

Waves of Possibility/Particles of Actuality

There is no dispute about the quantum facts—six decades of the most exotic and delicate experiments that human ingenuity could imagine. There is no dispute about the theory that accurately mirrors these facts: quantum theory has been exposed to possible falsification on a thousand different fronts and has perfectly passed every test that three generations of Nobel-hungry scientists could devise. The quantum reality controversy consists of the fact that scientists and philosophers have been unable to devise a single picture of the world consistent with both quantum theory and quantum fact. Physicists can perform quantum experiments of unprecedented accuracy and correctly predict the results of these experiments, but what they cannot do is clearly say "what is really going on" in these quantum experiments.

For example, quantum physics describes completely the behavior of atoms, a problem that baffled physicists of the last century. I have studied quantum physics for more than thirty years, but because of the quantum reality dilemma, I cannot tell my son what an atom looks like. Today nobody really knows what atoms look like. Ironically this inability to picture the atomic world does not arise because we know too little about atoms but because we know too much. As Werner Heisenberg, one of quantum theory's founding fathers, put it: "The conception of the objective reality of the elementary particles has evaporated in a curious way, not into the fog of some new, obscure, or not yet understood reality concept, but into the transparent clarity of a [new] mathematics."

The quantum reality problem arises from the fact that, more than sixty years after its inception, quantum physicists continue to represent an atom, or any other physical entity, not one way but two, depending on whether that atom is "being observed" or not. When an atom is being observed, the observer both sees it and describes it as possessing definite values for those attributes he chooses to look at, such as position, momentum, or spin. While it is being observed, the atom looks very much like a real object—a tiny little particle —one of the tangible building blocks of which the entire physical world is constructed.

A physicist observes the atom at a particular time, looks away for a moment, then observes it a second time. During both observations, the atom looks like a tiny object. However, if the physicist tries to describe the atom in between observations as a tiny object possessing definite attributes at all times, he finds that he cannot predict correctly the results of his second observation. On the other hand, if the physicist describes the unobserved atom in the peculiar quantum manner as a "wave of possibilities," he gets the right result every time for the second observation.

To get the right answer for his second look, a physicist is forced to describe the unobserved atom in a new and rather peculiar way—as a possibility wave, not as an actual object. What does it mean to represent an atom as a wave of possibilities? Instead of being located in one place like an ordinary object, the unobserved atom is represented, by a mathematical formula called the atom's *wave function*, as being in many possible places at the same time. (The conventional symbol for an entity's wave function is "Ψ," the twenty-third letter of the Greek alphabet.) In its mathematical representation at least, the unobserved atom seems to be everywhere and nowhere at the same time. The atom is everywhere because its wave function effectively spreads out over all space, although the wave's amplitude is largest near where the atom was last sighted. On the other hand, the unseen atom is "nowhere" because its wave function represents not the atom's actual presence but only the possibility of the atom's being in one particular place rather than another.

J. Robert Oppenheimer, the first director of the Princeton Institute for Advanced Studies, expressed the quantum physicist's descriptive dilemma this way: "If we ask, for instance, whether the position of the [unobserved] electron remains the same, we must say 'no'; if we ask whether the electron's position changes with time, we must say 'no'; if we ask whether the electron is at rest, we must say 'no'; if we ask whether it is in motion, we must say 'no.' " In the electron's wave function all of these activities are possibly present, but, until the electron is observed, none is singled out to be actually present. If we take this peculiar quantum wave function description seriously, then nothing "actually happens" between obsersations. What the math seems to say is that, between observations, the world exists not as a solid actuality but only as shimmering waves of possibility.

What does it mean to say that an atom's unobserved possibilities are wavelike? The possibilities for an atom to be in a particular location do not sit still but are continually vibrating at a particular frequency—so many cycles per second—a frequency that depends on the atom's energy content. In addition, when two of these oscillating possibilities come together, their amplitudes are added together, like sound or water waves, either to decrease (out-of-phase waves) or to augment (in-phase waves) the atom's chance of being in one place rather than another. In old-fashioned Newtonian physics, the possibility of something happening always increased when we increased the number of ways that it could occur. But for quantum possibilities, in the case of out-of-phase wave addition, the chances of something happening can actually be decreased by increasing the number of ways it can occur.

The quantum physicist treats the atom as a wave of oscillating possibilities as long as it is not observed. But whenever it is looked at, the atom stops vibrating and objectifies one of its many possibilities. Whenever someone chooses to look at it, the atom ceases its fuzzy dance and seems to "freeze" into a tiny object with definite attributes, only to dissolve once more into a quivering pool of possibilities as soon as the observer withdraws his attention from it. This apparent observer-induced change in an atom's mode of existence is called the *collapse of the wave function* or simply the *quantum jump*. One of the most fundamental unanswered questions in quantum theory is the nature of this quantum jump: does this drastic measurement-induced transformation of an atom's mode of being actually occur in the atom itself, or is the quantum jump a mere mathematical bookkeeping entry, representing the physicist's sudden increase in knowledge gained by observation? Does the quantum jump exist in the world as a real physical process or only in the physicist's mind: a mere mathematical fiction?

The price that the quantum physicist must pay to achieve his high-quality predictions is that he must train his mind to engage in a peculiar kind of quantum double-think: instead of a unified picture of nature, he must imagine the atom as many wavelike possibilities when not observed, as one particlelike actuality when observed. In light of the physicist's two-faced way of dealing with the world, the quantum reality question amounts to this: possibility waves are mathematical tools that serve the practical purpose of predicting experimental results, but, behind the theorist's tools and the experimentalist's results, what is the atom actually doing when we look at it and when we don't?

Eight Tentative Pictures of the Quantum World

At least eight different pictures of quantum reality have been proposed to explain (or evade) the question of what an atom is actually doing when nobody is looking at it, and to solve the so-called quantum measurement problem: the question of what actually goes on during a quantum jump. Here I will briefly summarize these eight positions. More detailed descriptions may be found in my recent book, *Quantum Reality: Beyond the New Physics*.

Quantum Reality 1: "There Is No Deep Reality"

First formulated by one of the most famous of the quantum pioneers, Danish physicist Niels Bohr, quantum reality 1 argues that only phenomena are "real." The *phenomena* are what we see before us, trees, rocks, stars, and the physicist's measurement instruments, Geiger counters, bubble chambers, and the unaided human senses. These things are undoubtedly real in every sense of the word. However, the atoms themselves are not so real. We know them only indirectly from the results of measurements. From these indirect and incomplete contacts with the atomic world, physicists have struggled to picture, like the blind men describing the elephant, what the atom looks like and have been utterly frustrated in their attempts to form an ordinary picture of that invisible world. In the late 1920s, Bohr took the position that the atomic world can never be pictured by human beings because it does not possess the same kind of actuality as trees, rocks, and stones. Atoms certainly exist, Bohr believed, but their mode of existence is of a type that cannot ever be grasped by human beings, who are constrained to live exclusively in the world of phenomena. Furthermore, our inability to picture atoms does not arise because we know too little about atoms but because we know too much.

Bohr's colleague, Werner Heisenberg, compared those physicists such as Einstein and Erwin Schrödinger who continued to search for ordinary pictures of the atomic world to believers in a flat earth: "The hope that new experiments will lead us back to objective events in space and time is about as well founded as the hope of discovering the end of the world in the unexplored regions of the Antarctic." Heisenberg's words have been remarkably prophetic: sixty years later we seem even further than ever from picturing the quantum world in the commonsense way envisioned by Einstein.

Quantum Reality 2: "Reality Is Created by Observation"

If only phenomena are real, then we are driven to ask, What is the essential nature of a phenomenon such as a tree, that distinguishes it from a less-real nonphenomenon such as an unobserved atom? Many physicists have concluded that "observation" is at the heart of every phenomenon. "No phenomenon is a real phenomenon until it is an observed phenomenon," quips quantum theorist John Wheeler, echoing the famous idealist Bishop Berkeley's slogan "Esse est percipi" (To be is to be perceived). Bishop Berkeley believed that nothing actually exists except as the perception of some being, that being we call "God" acting as "perceiver of last resort" whose constant attention keeps the world in existence whenever mortals close their eyes.

Concerning the question of nonhuman observation, Wheeler and most other physicists do not go as far as Berkeley: they do not believe that awareness, either human or divine, is necessary for making an observation. Rather an "observer" is anyone, or anything, that "makes a record." In their opinion, ordinary reality crystallizes out of some less real background substance in the form of "records"—collections of public, irreversible changes scattered throughout the natural world. The physicist's emphasis on the importance of observation (record making) in establishing the reality of quantum phenomena breathes new life into the old philosophical chestnut about whether a tree that falls unobserved in the forest makes a sound. The unprecedented success of this odd quantum way of dealing with the physical world has yanked the famous unobserved tree out of the philosophy classroom and rooted it at the heart of the most successful scientific theory ever known. Now not just naive college sophomores but sophisticated professional physicists are bedeviled by that lonely tree falling (or not falling?) in the vacant forest.

Together quantum realities 1 and 2 make up what is called the *Copenhagen interpretation* of quantum theory, after Niels

Bohr's hometown. To my mind this interpretation does not solve the quantum reality question so much as it evades it, by taking the real existence of macroscopic objects for granted and outlawing philosophical scrutiny of both the atomic world and its detailed interaction with measuring devices.

Quantum Reality 3: "Undivided Wholeness"

Old-fashioned Newtonian physics described the world as a collection of isolated particles interacting by means of "local force fields," such as gravity, electrical, and magnetic fields. A local field works according to the principle of *mediated interaction*: in order for a force, such as the earth's gravity, to affect a body, such as the moon, this force has to travel across the intervening space, with a velocity no greater than the speed of light. If the earth were suddenly destroyed by a comet, the moon would respond approximately 0.5 second later to the earth's absence. The opposite of a local force would be an interaction in which the earth could affect the moon directly, instantly and unmediated by an intervening field. Physicists from Galileo to Gell-Mann have always regarded nonlocal interactions of this sort as repugnant and in bad scientific taste. Isaac Newton once remarked that no philosopher in his right mind could imagine that such leap-frogging forces might exist in nature.

However, one of the most peculiar features of the quantum probability wave—the place where the quantum world differs most from classical expectations, according to Austrian physicist Erwin Schrödinger—is the fact that once two quantum systems have interacted, their possibility waves become entangled so that atom A's wave is mixed with atom B's wave in such a way that an action on atom A instantly and without mediation causes a change in atom B—in the mathematics, at least, if not in the world.

This sort of immediate, nonlocal interaction has no precedent in classical physics (where all interactions either occur through direct contact or are mediated by local fields), but it does resemble the belief in "contagious magic" of certain voodoo practitioners: the notion that something that was once a part of you, such as your hair or fingernail clippings, remains in direct contact, so that an action on the part instantly affects the whole. A recent discovery by Irish physicist John Stewart Bell sheds new light on the quantum entanglement process. Bell's theorem and its experimental verification by John Clauser (University of California at Berkeley) and Alain Aspect (University of Paris) prove that these nonlocal voodoolike connections not only are present in the mathematics but must exist as actual influences in the real world.

Bell's discovery that once any two atoms have interacted they remain really connected, their very beings entangled in a peculiarly intimate quantum manner, suggests that the best way to think about the quantum world might be not to imagine it as made up of separate parts in interaction but as some sort of undivided whole in which the "parts in interaction" picture arises as a simple approximation. The idea that the essence of the quantum world is an undivided wholeness was proposed by physicist David Bohm and others some time before Bell's discovery put speculations about quantum wholeness on a more substantial footing.

Quantum Reality 4: "The Many-Worlds Interpretation"

Hugh Everett, as a graduate student at Princeton, wished to use quantum theory to describe the whole universe. But because conventional quantum theory does not describe the world "as it is" but only as it appears to an observer, short of proposing an omniscient, omnipresent "observer of last resort" dwelling outside the material world—a somewhat unfashionable proposition among today's secular scientists—there is no way that quantum theory can be used to describe the universe itself. Instead, without changing the mathematics, Everett proposed a radical interpretation of the quantum formulation that reduced the role of the observer and immensely enlarged our view of what the word *universe* might mean.

Everett, who worked at the Pentagon on strategic planning until his untimely death in 1982, decided that the unobserved atom's quantum-possible positions were in fact actualities, not mere possibilities. The atom was actually in many places at the same time, but each of these atomic positions was located in a different universe. In Everett's interpretation, everything that can possibly happen does happen in one of the subuniverses of the grand Everett cosmos. We can envision the Everett cosmos as being made up of strands of spaghetti in spacetime, each strand a different possible history of what we would call the "whole universe" but which in fact is merely one subuniverse in a giant collection. Human observers dwell in many of these subuniverses, but they are not aware of the existence of their neighbors "next door." In Everett's model, quantum theory does not represent the probability of an event happening. All events happen in his world; none is left out. Rather, quantum theory represents for the observer the probability that he will find himself in universe A rather than universe B.

If the Everett interpretation gives a true picture of how the world actually works, then once again we have learned that ordinary human consciousness is a most inadequate tool for perceiving the world "as it really is." Einstein's special relativity theory, discussed in the previous chapter, describes the world as a changeless block of spacetime in which all events, past, present, and future, eternally coexist; this description does not jibe with everyday human experience of the world as a continually changing present moment. The physics-based worldviews of Everett and Einstein contradict our everyday experience: they both seem to be saying that the real world is immensely larger than what appears to our senses.

Of all quantum realities none is more outrageous than Everett's contention that myriad universes coexist with our own. However, because of its unified treatment of reality—no mysterious observer-created transitions from possibility to actuality in this model—Everett's extravagant vision has become increasingly popular among some quantum thinkers. Science fiction writers commonly invent parallel universes for the sake of a good story. Now quantum theory gives us solid motivation to take such stories seriously.

Quantum Reality 5: "Quantum Logic"

A small group of quantum thinkers believe that if the way that atoms possess their attributes cannot be expressed in ordinary language, then we should invent a new language more suitable for dealing with the quirky quantum world. But what is the smallest change that we can make in ordinary language in order to accommodate the strange quantum facts? What about preserving the words of our language but changing its "logic"?

Logic is the skeleton of our body of knowledge. Logic spells out the proper usage of some of the shortest and most important words in our language, words such as *and, or,* and *not.* In the midnineteenth century, George Boole, an Irish schoolteacher, in a book called *Laws of Thought,* invented an artificial symbolic language in which logical statements obeyed simple laws of arithmetic. Boole's codification of the rules of right reason laid bare the logical bones of ordinary language and founded the modern science of mathematical logic. Boolean logic has in modern times transcended its human roots: now this two-valued logical arithmetic forms the basis for the mechanical reasoning of computers.

Quantum physicists such as David Finkelstein at the University of Georgia remembered that Einstein had solved an important problem in physics—that of the nature of gravity —by introducing non-Euclidean geometry, the strange arithmetic of curved spacetime. Could it not be possible, these scientists asked, that the quantum dilemma might be solved in a similar way: by making a radical change in our very laws of thought? Instead of atoms whose positions are fuzzy until looked at, perhaps the world really consists of atoms whose positions are always definite (hence no "measurement problem") but we can only properly talk about these atomic positions by using a non-Boolean logic, involving new grammatical rules for combining the words *and, or,* and *not.*

The quantum logic approach does indeed solve some problems of quantum interpretation but leaves many others intact. Quantum logic seems at present to be still in its preliminary stages: a tentative proposition rather than a complete grammar of atomic behavior. My late friend Rockefeller University physicist Heinz Pagels criticized this approach by pointing out that if we accept quantum logic as the true logic of the world and somehow teach ourselves to think in this new way, then quantum mechanics becomes logical but the everyday world ceases to make sense. One of the biggest gaps in this ambitious project to non-Booleanize the world is the problem of how a world made of illogical atoms turns into our familiar world of ordinary logic when the number of such atoms becomes large.

Quantum Reality 6: Neorealism

Another attempt to resolve the measurement problem by imagining that atoms and other quantum entities always have definite positions whether observed or not is the "pilot wave" approach of French physicist-prince Louis de Broglie (who recently died at the age of 95) and British-American physicist David Bohm. Since the de Broglie–Bohm approach revives the notion of ordinary realism as a basis for quantum physics, I call this position "neorealism."

The main problem with a neorealist approach—bringing ordinary particles back into physics—is that the behavior of ordinary particles is just not crazy enough to explain the quantum facts. If atoms are really made up of ordinary particles, then some way must be found to make them behave as strangely as the quantum facts seem to demand. In the neorealist scheme, particles are ordinary and all of the world's quantum strangeness is relegated to an entity called the *pilot wave*. Unlike ordinary force fields such as gravity, which affects all particles within its range, the pilot wave acts on only one particle: each particle has a private pilot wave all its own that senses the location of every other particle in the universe. Although it extends everywhere and is itself affected by every particle in the universe, the pilot wave affects no other particle but its own. The pilot wave guides its private particle not by exerting forces but by supplying "information," like a radar beam. Furthermore, when a particle's personal pilot wave is actually calculated, it is immensely complex compared to the simpler conventional quantum description of its motion in terms of that particle's wave function.

Because this neorealist proposal does indeed rescue physics from mystical notions of particles that are not really there until you look, one might be tempted to accept the idea that every particle guides its journey through space via a personal radar wave. But two properties of the pilot wave are particularly unattractive to physicists and have hindered its easy acceptance.

Because it only affects one particle, the pilot wave is in principle unobservable. The existence and shape of pilot waves cannot be independently confirmed except indirectly as they each affect the motion of its associated particle. In addition, to supply its particle with accurately updated information about the whole universe, this wave must be able to transmit signals faster than light. Most physicists admire the ingenuity and philosophical simplicity of the neorealist approach but simply cannot stomach the notion that the world is permeated by 10^{80} complicated superluminal radar fields (one for each particle in the universe), not one of which can ever be observed.

Physicists do not like entities that are in principle unobservable: invisible pilot waves remind them of the equally invisible medieval angels dancing on the proverbial pinhead. Physicists are also uneasy about things that travel faster than light, since Einstein has shown that superluminal motions can be used to build time machines. Neorealists are quick to point out that the second objection is canceled by the first. If the pilot wave is unobservable, then its superluminal motions are unavailable for use in an Einstein time machine.

Quantum Reality 7: "Consciousness Creates Reality"

One of the most important intellectual figures of the twentieth century was Hungarian-born mathematician John von Neumann. In addition to his contributions to the field of pure mathematics, von Neumann initiated the study of economic and political behavior construed as rational games, devised the first theory of self-reproducing robots, and invented the notion of the stored-program computer. So fundamental were his contributions to the fledgling field of computer science that ordinary one-instruction-at-a-time computers—such as the machine this book was written on—are still referred to as "von Neumann machines."

In the early 1930s, von Neumann turned his restless mathematical mind to the newly developed physics of the quantum. Von Neumann put the loosely knit notions of Bohr and Heisenberg into rigorous form and settled quantum theory into an elegant mathematical home called *Hilbert space* where it resides to this day. (Unlike ordinary three-dimensional space, infinitely dimensioned Hilbert space is roomy enough to accommodate all of an atom's quantum possibilities at once.) In his magisterial tome *The Mathematical Foundations of Quantum Mechanics*, regarded by many scientists as "the bible of quantum theory," von Neumann exposed and boldly attacked the formidable quantum measurement problem, which most physicists had been too complacent or intimidated to confront.

In his "quantum bible," von Neumann objected to the Copenhagen practice of dividing the world into two parts: quantum entities (possibility waves) and classical measuring instruments (actual objects possessing definite attributes). Von Neumann believed that Bohr's followers were wrong to divide the world into two fundamentally different parts. Our world is whole, not split in two, claimed von Neumann. It possesses a single nature, and that nature is certainly not classical. However, if the world is entirely quantum-mechanical, as von Neumann believed, the theory of the quantum unequivocably requires that it be described in terms of possibility waves, not as a collection of actual objects possessing at all times a definite value for each of their physical attributes. A totally quantum world is a world of pure possibility. Nothing ever really happens there; everything just hesitates forever on the brink of actuality. Compared to the actual world—the old-fashioned, definite "yes or no" world of classical physics— the quantum world resembles a fairy-tale land built solely of ambiguous "maybes."

To resolve the measurement problem in von Neumann's all-quantum world, something new must be added to "collapse the wave function," something that is capable of turning fuzzy quantum possibilities into definite actualities. But since von Neumann is forced to describe the entire physical world as possibilities, the process that turns some of these maybes into actual facts cannot be a physical process. To collapse the wave function some new (actual not possible) process must enter the world from outside physics. Searching his mind for an appropriate actually existing nonphysical entity that could collapse the wave function, von Neumann reluctantly concluded that the only known entity fit for this task was consciousness. In von Neumann's interpretation, the world remains everywhere in a state of pure possibility except where some conscious mind decides to promote a portion of the world from its usual state of indefiniteness into a condition of actual existence.

Von Neumann's position (based on physics) is very close to Bishop Berkeley's (based on theology): nothing in this world is real unless it is being perceived by some mind. "All those bodies which compose the mighty frame of the world," said the Irish bishop, "have no substance without a mind." As a professional mathematician, von Neumann was accustomed to following his logical arguments boldly wherever they might lead. Here, however, was a severe test for his professionalism, for his logic leads to a particularly bizarre conclusion: that by itself the physical world is not fully real, but takes shape only as a result of the acts of numerous centers of consciousness.

Ironically this conclusion comes not from some otherworldly mystic examining the depths of his mind in private meditation, but from one of the world's most practical mathematicians deducing the logical consequences of a highly successful and purely materialistic model of the world—the theoretical basis for the billion-dollar computer industry.

Quantum Reality 8: The Duplex World of Werner Heisenberg

No one was more aware of the conceptual difficulties involved in attempting to describe the state of an unobserved atom than Werner Heisenberg, the Christopher Columbus of the new quantum world, who discovered in 1925 the first successful mathematical theory of the quantum. Modern quantum theory has immensely elaborated Heisenberg's initial insight, but, despite an explosion of new experimental results, the essence of the theory has not changed in the intervening years: the philosophical difficulties that troubled Heisenberg and his colleagues are with us to this day. "The problems of language here are really serious," Heisenberg remarked. "We wish to speak in some way about the structure of the atoms and not only about the 'facts'—for instance, the water droplets in a cloud chamber. But we cannot speak about the atoms in ordinary language." Niels Bohr and his Copenhagen colleagues had convinced most physicists that it was humanly impossible to form pictures of the atomic world. Swimming against the physics mainstream, Heisenberg courageously took up the challenge of how to express the quantum behavior of atoms in ordinary language.

Heisenberg constructed his picture of reality by taking quantum theory seriously, not merely as a device for calculating experimental results but as a true picture of the world. Heisenberg proposed that, since quantum theory represents the unobserved world as possibility waves, then perhaps the world might really exist, when not looked at, as mere waves of possibility.

According to Heisenberg's scheme, there is no deep

reality—nothing down there that's real in the same sense as the phenomenal facts are real. The unobserved world is merely semireal and achieves full reality status only during the act of observation:

> In the experiments about atomic events we have to do with things and facts, with phenomena that are just as real as any phenomena in daily life. But the atoms and the elementary particles themselves are not as real; they form a world of potentialities or possibilities rather than one of things or facts. . . .
>
> The probability wave . . . means a tendency for something. It's a quantitative version of the old concept of potentia in Aristotle's philosophy. It introduces something standing in the middle between the idea of an event and the actual event, a strange kind of physical reality just in the middle between possibility and reality.

One of the inescapable facts of life is that all of our choices are real choices. Taking one path means forsaking all others. Ordinary human experience does not encompass many contradictory events all happening at the same time. For us the world possesses a concreteness and uniqueness apparently absent in the atomic realm. Only one event at a time happens here, but that one event really does happen.

The quantum world, on the other hand, is not a world of actual events like our own but a world teeming with numerous unrealized tendencies for action. These tendencies are continually on the move, growing, merging, and disappearing according to exact quantum laws of motion discovered by Heisenberg and his colleagues. But despite all this frantic atomic activity nothing ever really happens down there. As long as they remain unobserved, events in the atomic world remain strictly in the realm of possibility.

Heisenberg's two worlds are bridged by a special interaction that physicists call a "measurement." During the magic measurement act, one quantum possibility is singled out; abandons its half-real, shadowy sisters; and surfaces in our ordinary world as an actual event. Everything that happens in our world arises out of possibilities prepared for us in that other —the world of quantum potentia. In turn, our world sets limits on how far pools of potentia are permitted to spread. Because certain facts have become actual in our world, not everything is equally possible in the quantum world. In Heisenberg's duplex vision, there is no deep reality, no deep reality-as-we-know-it. Instead the unobserved universe is made up of possibilities, tendencies, urges. Our solid everyday world is founded, according to Heisenberg, on something no more substantial than a promise.

One of the major unsolved problems of the nineteenth century was the so-called ultraviolet catastrophe. If atoms were miniature solar systems that obeyed the rules of classical physics, then they should explode in a burst of bright purple light after about a billionth of a second. Quantum theory resolved this problem of atomic stability—the quantum atom left to itself is essentially immortal—but raised new problems of its own. Classical physics could not make a universe that would last more than a billionth of a second. But quantum mechanics—in the von Neumann picture, for example—cannot make a real universe at all, or, at least, not without some outside help from nonmaterial forces. Quantum theory solved the ultraviolet catastrophe—the totally incorrect prediction that the lifetime of a classical universe is only a billionth of a second—but replaced it with its own "existential catastrophe." In the quantum world, the fact that the universe exists at all, as actual fact not mere possibility, is not completely explained. The Old Physics could not make the universe exist for more than a fraction of a second. However, for that instant, it really did exist. On the other hand, for the New Physics, the fact that the universe exists at all is somewhat problematic.

The quantum "existential catastrophe" differs in one important way from the ultraviolet (UV) catastrophe. The UV catastrophe predicted something that one might actually

observe—the explosion of the universe—but the quantum catastrophe mainly involves the real nature of unobserved atoms, something that we can never, in principle, ever observe.
The quantum reality problem is, strictly speaking, not a physics question at all, but a problem in metaphysics, concerned as
it is not with explaining phenomena but with speculating about
what kind of being lies behind and supports the phenomena.

It should be mentioned that each of these eight realities
from Bohm's neorealist particle-plus-wave model to von Neumann's consciousness-created world is perfectly compatible
with the same quantum facts. We cannot use experiments—
or at least experiments of the usual kind—to decide among
these conflicting pictures of what lies behind the phenomenal
world.

However, this lack of experimental verification does not
render these quantum realities useless. One of the most important uses for metaphysical pictures is to help extend quantum physics into new areas: models of mind, for instance.
Without tentative models of what is really going on in the
world, quantum theory remains nothing but opaque mathematical formalism, a very sophisticated kind of ignorance. By
itself, without interpretation, the mathematical formulas resemble a magic spell that works every time: to exert his power
over the world the magician (mathematician) who uses the
spell never has to know why it works. For the purpose of
exploitation, the mathematics alone suffices, but for the purpose of exploration even a bad picture of what is going on may
lead to new discoveries. The investigator of new realms might
regard these eight quantum realities as tentative maps of the
borders of an unknown territory: the whole universe as it actually exists, of which physical reality is just one part among
many.

For the construction of models of mind and clues to the
true role of consciousness in the universe-as-a-whole, these
eight quantum realities (with two exceptions) offer tantalizing
suggestions. The two realities least friendly to theories of consciousness are, in my opinion, the quantum logic option (quan-

tum reality 5) and the neorealist picture of the world (quantum reality 7). The quantum logicians seem to believe that the quantum reality problem is merely linguistic and can be fixed simply by adopting a new language. The neorealists hope to return to a clearly visualizable world made up of ordinary particles and not-so-ordinary waves. Both of these pictures are completely self-sufficient—need no new elements from outside physics—and compatible with a purely materialistic universe. As with old-fashioned Newtonian physics, these quantum pictures leave no room, no role for mind to play in the world.

On the other hand, the Copenhagen picture (quantum realities 1 and 2) holds that the unobserved world that sustains this one is not ordinary, and that the act of observation drastically modifies this strange substratum, changing it at every moment into the world of the everyday. Heisenberg's picture (quantum reality 8) attempts to say more about the deep substratum: it is made of tendencies, of possibilities, not actualities. The quantum wholeness picture (quantum reality 3) adds to Heisenberg's specifications the notion that the substratum's "parts" are intimately linked together in a particularly quantum way. Von Neumann extends the Copenhagen picture by revealing more about the mysterious measurement process: a measurement only happens in some mind, he says. Von Neumann's hypothesis not only makes room for mind but gives it an independent role to play in constructing the phenomenal world. Von Neumann's model of reality treats mind as "elemental," as fundamental as quarks and gluons for the proper functioning of the universe. Lastly, the many-worlds picture (quantum reality 4) suggests that our human experience is part of a larger experience enjoyed by similar beings—our other selves—in similar universes quite near by (near by in Hilbert space, that is).

We saw in the last chapter that the raw material of Culbertsonian consciousness consists of tangles of world lines in four-dimensional spacetime. The substratum of most quantum theories of consciousness is Heisenberg's picture of the material world, consisting not of actual facts but of unrealized possibilities for existence. Mind (or minds) then brings the factual world into existence by selectively realizing some of these possibilities at the expense of others. In the next chapter we will examine some important features of quantum theory and how it has been combined with our knowledge of brain function to build tentative models of how inner experience of the human kind might shape and be shaped by matter operating according to these strange quantum rules.

quantum quintessence: randomness, thinglessness, inseparability

If a person does not feel shocked when he first encounters quantum theory, he has not understood a word of it.

—NIELS BOHR

The mathematics of quantum theory yield results that coincide with experimental findings. That is the reason we use quantum theory. That quantum theory fits experiment is what validates the theory, but why experiment should give such peculiar results is a mystery. This is the shock to which Bohr referred.

—MARVIN CHESTER

Scene: Rudi's artificial consciousness lab at Pleasure Dome University. Located in the basement of one of the campus's oldest buildings, the lab is full of outdated electromechanical equipment tended by obsolete robot lab assistants. The lab's most striking feature is a shelf full of humanoid heads, some of whose eyes follow you as you move about the room. Most of these heads have thick cables running out of their necks to various types of computerlike machinery. Rudi is directing the assembly of a new head as he talks to Nick.

RUDI: You can't imagine how hard it is to get funding for research on artificial consciousness these days. The main trouble is that we've been working for years without a single experimental breakthrough. Lots of theorists have come up with possible models of consciousness, but most of them are too vague to be tested, and those that we've been able to build hardware around—none of those has ever worked.

NICK: What would it mean for an artificial consciousness model actually to "work"?

RUDI: That's a good question. Since consciousness as we define it is an experience not a type of behavior, a good awareness module would have to be able to produce observable experiences in some being. But we don't yet have a mind link that allows us to share the private experiences of other beings. So to test an alleged awareness module, the experimenter himself must act as his own subject. You can't imagine all the crazy devices I've hooked myself up to over the years. Look here! I've got a 64-pin biojack at the base of my brain for direct access to my reticular formation and points north.

NICK: If none of these awareness modules works, why are you experimenting with Claire's head? Why are you building up her hopes to become a conscious being?

RUDI: Well, it's not exactly true that none of these modules works. Some of them produce quite jazzy experiences—a kind of electric psychedelic. But it's difficult to know whether these devices are actually producing inner experiences or just changing the inner experiences that I already have. To do this thing right, you should take a being without consciousness and see whether the module gives it a flow of inner experiences. So I've been taking some of my trippiest modules to articulate, sophisticated robots, and offering to hook them up. I suspect that the difference between a conscious and an unconscious robot would be so great that the robot's report, although secondhand, will be convincing evidence that the device worked. Claire is going to test one of my "Eccles gate modules" this weekend. An *Eccles gate* is a neural net whose connections are made with quantum synapses. The experiences it produces in my mind are really "far out," as our ancient forerunners in the mind expansion business used to say. If you had a brain jack, I'd let you experience the Eccles gate for yourself.

NICK: No thanks, Rudi. Just tell me in words what the experience is like. Does an Eccles gate help your mind to think quantum-logically?

RUDI: It's pleasant and terrifying at the same time. It's like floating in a warm sea of expanded mental and bodily options. Everything is intense and at the same time unfocused, if that makes sense. In the Eccles gate experience, my mind seems to be immersed in a borderless ocean of absolutely certain uncertainty.

NICK: Well, that sounds pretty "quantum" to me. What's so frightening about it?

RUDI: One of the greatest uncertainties I experience is my own identity—a classic case of ego loss. The Eccles gate experience, as intense as it is, seems to have no center. Something big is undoubtedly happening but there's nobody that it's happening to.

NICK: But isn't the unity of human consciousness one of the most important features of ordinary awareness? Seems to me that you're moving in the wrong direction if these quantum devices dissolve the mind's essential unity.

RUDI: Our ordinary notions of mental unity may be somewhat naive. I think we've been hypnotized by centuries of Newtonian thinking to believe that our minds are just machines for manipulating things called "experiences" and that one of these experiences is a thing called "self."

NICK: If an experience isn't a thing, what is it then, a process?

RUDI: No, a process is also a thing—just a bit more complicated. To understand quantum consciousness—which I believe is the same as ordinary awareness—we must find a way to go beyond "thing-thinking."

NICK: Sounds nice, Rudi. But how do you plan to do that? Are you going to rewire your brain to think quantum-logically?

RUDI: No, part of the brain is already a quantum device—the part that's conscious. All we have to do is ignore the vast amount of Newtonian data processing that goes on in there and pay close attention to how raw consciousness really feels from the inside. But to do this without Newtonian preconceptions, that's the hard part.

NICK: Sounds like you think that mind scientists ought to sit in meditation like Buddhist monks.

RUDI: Right. You can't really study consciousness without paying close attention to what consciousness actually feels like. Know thyself, big-brain lab monkeys.

Legend has it that Isaac Newton, on leave in 1665 from plague-ridden Cambridge University, was inspired to discover the principle of general gravitation by the sight of a falling apple in his mother's garden. The same force that pulls the apple must also pull the moon, he guessed. After connecting the fall of his mother's apple with the moon's orbit, so the story goes, Newton went on to elucidate the mechanics of the universe, showing that the particles that compose the world move, not by divine whim, but according to the dictates of universal, impersonal, mathematical laws. Controlled at every level by unchanging, deterministic laws of motion, the physical world of Newton and his followers came to resemble a giant machine, a cosmic clockwork whose every action was preordained, completely predictable from its initial state at the moment of creation.

Visualizing the universe as one giant clock captures the main features of the Newtonian worldview: the world is made up of objects (the clock's gears and bearings), moved by impersonal forces (the clock's mainspring), subject to deterministic laws of motion—when something is utterly reliable, we say "it runs like clockwork."

Pavlov's Dog

Another image of the Newtonian worldview, more relevant to models of mind, is that of Pavlov's dog. In the beginning of this century, Russian physiologist Ivan Pavlov trained dogs to associate arbitrary signals such as the sound of a bell or the sight of a card with a circle printed on it with the presentation of food. The dogs soon learned to salivate for the symbols alone in the absence of food. Pavlov's *conditioned response*—the predictable connection between an animal's physiological response and some external stimulus—became the cornerstone of the science of behavioristic psychology. Behaviorism extended the Newtonian worldview to the realm of living beings, treating dogs as well as human beings as clockwork creatures, whose behavior could be described solely in terms of stimulus/response reactions, without regard to their inner experiences. In this "Newtonian psychology," Pavlov's dog, and by extension Ivan Pavlov himself and the rest of us, is a mere machine, utterly predictable once experimental psychologists learn the underlying laws of behavior, animal "laws of motion" corresponding to Newton's physical laws.

Like Pavlov's dog, the non-Newtonian quantum worldview also has its animal mascot—Schrödinger's cat. More than any other image, the puzzling status of Schrödinger's fuzzy feline symbolizes the strange condition in which every quantum object must dwell: a condition described as vibrating possibilities when not looked at, as solid actualities when observed. Schrödinger's cat is a large-scale quantum object: our animal stand-in, for whom the paradoxical quantum mode of existence is allegedly a matter of direct experience.

Schrödinger's Cat

Erwin Schrödinger, a professor at the University of Vienna and discoverer of *Schrödinger's equation*, the basic law that governs the motion of quantum possibility waves, was profoundly distressed by the notion that, in some sense, the quantum world does not fully exist until it's observed. One might relieve some of this distress, Schrödinger reasoned, if the quantum reality crisis could be securely confined to the world of atoms, which are anyway too small to see. The "degree of unreality" of an object, the range of its Heisenberg uncertainty, is measured by a number called *Planck's constant* after German physicist Max Planck. For objects as small as atoms, the Heisenberg uncertainty of its surrounding electrons is as large as the atom itself, but for ordinary objects like bricks and bathtubs, this mite of unreality is inconceivably small, like the glow of a firefly compared to the glare of the sun. The minuscule size of Planck's constant, compared to that of ordinary motions, is responsible for the enormous success of Newtonian physics and the fact that the everyday world seems quite ordinary despite its weird quantum underpinnings. Because Planck's constant is so tiny, quantum effects are much too small to be noticed under ordinary circumstances. As we shall see, one of the main obstacles to a quantum theory of mind is the smallness of Planck's constant: where in the brain is there a system so tiny that quantum uncertainties dominate its operation? The strange case of Schrödinger's cat may suggest an answer to this question.

For those who had hoped that quantum strangeness might be permanently confined to the scale of atomic objects, Schrödinger, in 1935, cooked up an unpleasant surprise—a thought experiment based on quantum ideas that continues to trouble physicists to this day. Schrödinger began by devising a situation in which the small quantum uncertainty of a single system could be split into two parts, so arranged that each of these parts leads to radically different experimental consequences.

For instance, imagine a single photon of light impinging on a half-silvered mirror of the type used in the windows of modern office buildings. According to quantum theory, if this photon is not observed, its possibility wave splits at the mirror surface and takes both paths, one-half of the wave going through the mirror and one-half bouncing off. Now arrange two *photon detectors*—devices that produce an electrical signal in response to light—in each of these two paths. Until these detectors are observed, quantum theory describes them both as possibility waves, each triggered by its own photon possibility wave. Now place this device (mirror plus detectors) inside a box, made soundproof and lightproof to ensure that observation of its contents is impossible, so that, immune from external observation, whatever is inside the box will remain (according to quantum theory) in a state of pure possibility.

Inside the box, continues Schrödinger, we have also placed a cat. The cat is fed if detector 1 puts out a signal; the cat is killed if detector 2 fires. (The fed/dead distinction was chosen for dramatic effect. Schrödinger actually liked cats and referred to his quantum cat box as a "hellish device.") Now as long as the box is not opened, if we take quantum theory seriously, we must describe the cat as being both alive and dead at the same time. "This does not mean the cat is sick," adds my friend Bruce Rosenblum, a physicist at the University of California at Santa Cruz. The cat-in-the-box is in a new state impossible to imagine in commonsense terms, the kind of state that atoms are almost always in (except when they're being looked at), the kind of paradoxical state of multiple unrealized possibilities that according to quantum theory must underlie the entire physical world.

What happens when we open the cat box and attempt to verify the cat's alleged twofold mode of existence? Just what you might expect. Just as you look at it, the cat jumps from a state of possibility to a state of actuality. Whenever you look, the cat is always found to be either alive or dead, never both. This is what quantum theory predicts would happen, and we would not be surprised to find either of these outcomes to Schrödinger's hellish thought experiment. The real question is this: Before you open the box, what is the true condition of Schrödinger's cat?

Schrödinger believed that the notion of an alive/dead cat was patently absurd. If the quantum theory as it existed in

1935 leads to such a conclusion, then this theory must be wrong. However, more than fifty years later, after passing hundreds of tests Schrödinger could never have imagined, today's quantum theory is stronger than ever and continues to predict that unobserved cats of the Schrödinger variety must be both alive and dead at the same time.

Other scientists have proposed that even inside a sealed box, the photon detectors are "making observations" because of the "thermodynamically irreversible processes" (record-making events) that occur inside them. However, such scientists were accused by mathematician John von Neumann of not following their own rules. According to quantum theory, any unlooked-at object, even a record-making device, must be described not as a fixed actuality but as a sheaf of possibility waves. Show me, challenged von Neumann, what is intrinsically different about a record-making event that would exempt it from this quantum rule.

Others have pointed out that, even if von Neumann is right to say that unobserved photon detectors exist only as possibilities, the cat must know its own state. Whatever the condition of unlooked-at inanimate objects, the cat is certainly fully qualified to observe itself and establish a condition of actuality out of the fuzzy potentialities created in its box by the photon hitting the half-silvered mirror. This solution raises the question of the degree of consciousness that various animals possess. If Schrödinger's cat possesses enough self-awareness to "collapse the wave function," what about "Schrödinger's amoeba"?

One could imagine bypassing the question of animal consciousness by substituting an inanimate object for the cat. Make the box stronger, suggested Einstein. Replace the cat by a stick of dynamite that would be detonated or not depending on which photon detector was triggered. Now if quantum mechanics is correct, Einstein's box contains both a loud explosion and a quiet dud at the same time. What is the sound of one (unobserved, quantum) bomb flapping? Schrödinger's cat and Einstein's bomb are two different dramatizations of the basic quantum reality question: What is the existential status of unobserved quantum objects? The most important feature of these thought experiments is that Schrödinger and Einstein have taken the quantum reality question out of the humanly inaccessible realm of atoms and molecules and placed it squarely in the ordinary world of cats and high explosives.

The example of Schrödinger's cat offers a picturesque way to express the two major philosophical problems presented by quantum theory: the quantum reality question (How can we adequately conceptualize the unobserved world?) and the quantum measurement problem (How does the observed world emerge from the unobserved background?). Stripped of philosophical jargon, the gist of these problems can be stated: What really happened in that box to Schrödinger's cat? and How did Schrödinger's cat turn into Pavlov's dog? Until he can give convincing answers to these two questions, no physicist can really claim to understand quantum theory.

The basis for most quantum theories of consciousness is that mind enters the material world via the leeway afforded by Heisenberg's uncertainty principle. To the extent that matter is uncertain, mind can have a say in the motion of matter by selecting which quantum possibilities are realized. However, in almost all cases the range of quantum uncertainty is exceedingly small (we will estimate it later) compared to the size of motions in large systems such as the brain. The case of Schrödinger's cat shows, however, a possible way for a quantum mind to exert a large effect. One needs to look for the biological equivalent of a half-silvered mirror, a physical system that splits quantum possibility waves into two or more components each of which leads to a radically different outcome. Such possibility-wave splitters might be called "quantum razors." A mind that could control the output of a quantum razor would be able to produce effects in the material world entirely out of proportion to the tiny range of motions allowed by the Heisenberg uncertainty principle. In the case of Schrödinger's cat, the razor-sharp decision of such a mind amounts to a matter of life or death.

In the von Neumann interpretation of quantum theory (quantum reality 7), consciousness is a process lying outside the laws that govern the material world. It is just this immunity from the quantum rules that allows mind to turn possibility into actuality. Because quantum-based minds are inevitably different in substance from the matter they control, theories of such minds are bound to be dualistic. Humans, animals, and conscious robots have "ghosts in their machines," as Gilbert Ryle's scornful description of dualism would have it. If we are all ghosts inhabiting quantum machines, souls out for a spin in matter-made automobiles, what are the chief features of our vehicles? What can quantum theory tell us about how a ride in such a machine might feel?

The quantum world differs in many ways from the Newtonian clockwork cosmos that it supersedes. Three features of the quantum world seem to me to be particularly important for models of mind: randomness, thinglessness, and inseparability. Coincidentally, it was just these three features that profoundly disturbed Einstein. Einstein was impressed by the success of quantum theory but could not accept the notion that at its core the world is random, is not made of things, and is connected in a peculiar way that seems to defy common sense and his own theory of relativity. One of the best ways to look at quantum theory, I believe, is through the three windows of Einstein's pet peeves.

Quantum Randomness

Before an atom is looked at, physicists describe it as waves of possibility—a superposition of many possible atomic actions at once, with a range of variation set by Heisenberg's uncertainty principle. When the atom is observed, one of these possibilities becomes real, but quantum theory gives no indication which one will be actualized: it appears to be a matter of pure chance.

One example of quantum randomness is a radioactive isotope A that decays into isotope B with, say, a half-life of 10

minutes. Prepare a million such radioactive atoms at 10:00 P.M.
By any of the physical methods known to science, these atoms
all appear to be identical. Yet at 10:10 P.M. half of them have
decayed into isotope B; half of them are unchanged. There is
no way of predicting which atoms will "die" and which will
"live" during the first 10 minutes. By 10:20 P.M. half of the
remaining A-type atoms will have transformed into isotope B.
The half-life of these atoms is highly predictable but the
precise time at which an individual atom decides to decay is
completely unpredictable.

In a fluorescent lamp, mercury atoms are excited by electron collision into states of higher energy. These atoms return
to their ground states in a fraction of a second by emitting a
photon of light. The time of emission, the direction, and the
polarization of the emitted photon are completely quantum-
random, unpredictable by present scientific means.

Schrödinger's cat represents a third example of quantum
randomness. When someone opens the box and looks inside,
the cat goes from being in a sum of a possibly live cat plus a
possibly dead cat to one actual cat, either fully alive or completely dead. Which cat possibility is actually realized is not
predictable within the quantum theory. The outcome of the
cat-in-the-box experiment is completely random.

Einstein could not accept the notion that atomic events
were totally uncaused—that at this level of phenomena physicists must give up searching for explanations because physics
stops here. "I cannot believe that God would play dice with
the Universe," he said. However, Rockefeller University physicist Heinz Pagels had the opposite view of quantum randomness.

If you want to build a robust universe, one that will
never go wrong, then you don't want to build it like
a clock, for the smallest bit of grit could cause it to
go awry. However, if things at base are utterly random, nothing can make them more disordered. Complete randomness at the heart of things is the most

stable situation imaginable—a divinely clever way to
build a sturdy universe.

(Or as Santa Cruz poet Greg Keith says: "When you stand on
random, you can't fall much.")

However admirable as a scheme for producing a stable
unconscious universe, ultimate randomness on the face of it
seems an unsuitable ground for elemental mind. A consciousness whose decisions were completely random would be no
freer and certainly less dependable than a mind made of ultra-
reliable Newtonian clockwork.

One way of describing quantum randomness is to say that
we cannot predict which possibility will actualize. Another is
to say that identical situations can give rise to different outcomes. In the Newtonian world, identical situations always led
to identical outcomes, but in the quantum realm, two atoms
physically identical in every possible way can exhibit very different styles of behaviors.

A third way of conceptualizing quantum randomness is to
say that the causes, if any, of atomic behavior do not lie in the
physical world: no amount of physical examination will ever
allow us to predict exactly what an atom will do next. This
way of expressing quantum randomness is especially condu-
cive to models of consciousness for it opens up the possibility
that the ultimate cause of material phenomena is not material
at all but stems from an essentially mental realm.

If mind exerts its power over nature by selecting which
quantum outcome actually occurs, then our perceived freedom
of action is not illusory, for physics as currently conceived regards quantum events as essentially uncaused, unrestrained
by prior physical events. Although each quantum event is completely causeless, the pattern of quantum events as a whole is
constrained statistically to follow the pattern of possibilities
contained in the system's wave function. The pattern of quantum events is precisely predictable but each individual event
is not. In this regard quantum possibilities resemble classical
probabilities such as those for a dice game. Each throw of

the dice is unpredictable (in a fair game) but over the long run, a pattern emerges that favors "sevens" over "elevens" or "twos." Random processes paradoxically are not lawless but must obey rules too. However, what might be called "Robert's Rules of Random Order" govern the behavior of large numbers of events but not the individual events that make up these aggregates.

Exerting free choice modified by statistical constraints may be compared with the act of speaking or writing a language. We feel free to say whatever we please, but when our utterances are examined by mathematicians, the statistical distribution of letters and spaces is remarkably stable and independent of the sense of the utterances. Heinz Pagels regarded quantum theory as the "cosmic code," the language in which nature has chosen to express herself. Is it possible to take this metaphor literally and regard the world as built up of immaterial minds, big and little, that actually "speak" the material world into existence? The possibility waves on which quantum science is based, in this view, would play a decidedly minor role in the world's affairs. Quantum possibilities would be merely the speech statistics of a vast cosmic utterance: the universe's real meaning would lie elsewhere, in the immaterial communication intentions of elemental minds.

Quantum Thinglessness

A thing is any entity that possesses definite attributes whether looked at or not. Not only do quantum objects such as atoms or Schrödinger's famous two-valued cat possess no attributes in their unlooked-at state, but the attributes that they acquire in the act of being observed depend to some extent on how they happen to be observed. Einstein could not accept the notion that quantum objects—atoms, electrons, photons, or boxed cats—do not possess attributes of their own, but acquire them only in the act of observation. "I cannot imagine," he said, "that a mouse could drastically change the

Universe by merely looking at it." The belief in an external world independent of the perceiving subject, Einstein maintained, is the basis of all natural science. "Atoms are not things," retorted Werner Heisenberg. Cats-in-boxes are not things either, according to the orthodox quantum view. Until Einstein's mouse, or some other observer, meets Schrödinger's cat, looks her over, and endows her with definite attributes, the cat-in-the-box must be regarded as a nothing, as a mere superposition of quantum possibility waves.

Besides the mysterious transition between possible and actual, an additional feature of quantum thinglessness is the necessary existence of pairs of incompatible attributes. In Newton's physics, any physical attribute could be measured in principle to any degree of precision regardless of what other attributes were measured at the same time. Quantum attributes are different. For reasons related to the wave nature of the quantum description, quantum attributes always exist in pairs: every attribute A has at least one partner (some have more)—attribute B—to which it is inextricably linked. If one chooses to measure attribute A precisely, then one must forgo measuring attribute B. Since the measurement process, in the usual quantum picture, is viewed as turning possibilities into actualities, the observer's choice of what to measure (attribute A, for instance, rather than attribute B) amounts to a choice of which attribute shall enter the world of reality and which shall remain unactualized. The limitation on observation imposed by the existence of quantum-incompatible attributes, a limitation not present in classical physics, confers as a kind of compensation a certain power on the observer, a power likewise not present in classical physics, the power to decide what sorts of attributes an object will seem to possess. Hence, in a certain sense, the existence of incompatible attributes—not all of which can be realized at the same time—makes the observer a co-creator of reality along with nature. Let's see how this minor sort of reality creation actually works.

For the physicist, the most familiar pair of incompatible

attributes are position and momentum. Suppose an electron (an undeniably quantum object) is approaching you, yearning to be measured. You decide to measure its position by deploying a position meter in its path. The electron will then appear to acquire some definite position. (Exactly where the electron will appear is unpredictable, because of quantum randomness.)

However, you could have decided to measure the electron's momentum, by inserting a momentum meter in its path. Then the electron will appear to acquire a definite (unpredictable) momentum, at the price of forgoing any knowledge concerning its position. By deciding what attribute you want to measure and deploying the appropriate instrument, you invite that attribute, but not its partner attribute, to manifest itself in the actual world. In the unobserved world of pure possibility, incompatible attributes can exist without contradiction, but there is room in the world of actuality for only one partner. Which partner appears is not specified in the quantum description but is decided by the type of measurement the mouse (or other competent observer) decides to make.

Examples of incompatible attributes include the position/momentum pair already mentioned and polarized light: the light from an excited mercury atom is polarized in a Schrödinger cat type of way—in a superposition of classical polarization possibilities. One can measure the plane-polarization attribute of a mercury photon or the circular-polarization attribute, but not both. How you choose to measure mercury light is a vote for which kind of polarization will emerge into this world (from the indecisive world of the unobserved).

A third example is Schrödinger's cat. In the standard example, the observer chooses to ascertain the live/dead condition, to observe what we might call the cat's "mortality attribute." But, like mercury light waves, cat possibility waves can be analyzed in more ways than one.

The mathematics of the cat situation allow us to imagine an attribute incompatible with mortality. Call this "attribute X." Like the live/dead situation, attribute X consists of two outcomes, which I will call "plus cat" and "minus cat." The possibility wave that describes the plus cat option is obtained by adding the live and the dead cat wave in phase; the minus cat wave results from adding the live wave to the dead wave in an out-of-phase manner. Such manipulations make no sense in classical physics, where attributes are fixed once and for all. Like the mortality attribute, attribute X is a perfectly good quantum observable. However, at this stage in our understanding, no physicist knows how to make a device that will measure attribute X, nor can he tell you what a plus cat might look like. So although Einstein's mouse can in principle measure the cat-in-the-box's X attribute, it is unlikely that he has the experimental know-how to carry out this sort of measurement. However, if a mouse did somehow learn to make the X-attribute measurement, he would see, upon opening the box, either a plus cat or a minus cat. Nobody can predict which it would be.

This freedom to choose which attributes a quantum system will manifest may be compared to the choice of a game that can be played with a given deck of cards. The dealer chooses the game, then "Lady Luck" selects which cards are dealt. In the quantum measurement situation, the observer selects the kind of attribute he wants to look at, then quantum randomness selects the particular value of that attribute that the observer actually sees.

Quantum mind scientists will certainly have to incorporate quantum thinglessness in their models, when they try to describe the details of how mind causes matter to come into being. Somewhere at the mind/matter interface, the mind must not only select how the dice fall but call out the name of the dice game as well.

Deciding which attribute to bring into existence involves more than merely making up your mind. Actually to perform an observation, you must provide an appropriate physical context that will invoke the particular attribute that you intend to observe. Different physical contexts—different types of measuring devices—must be deployed to measure position rather than momentum, the cat's X attribute rather than the cat's mortality attribute. One important facet of the quantum measurement problem—how Schrödinger's cat turns into Pavlov's dog—is how definite measurement contexts are established "in the wild." If the construction of simple quantum events from raw potentia is problematic, even more puzzling is how definite measurement contexts emerge out of mere possibilities. How does nature decide—and make her decisions stick—whether to manifest an electron's position rather than its momentum?

A quantum mind faces the same measurement problem when it desires to manifest some aspect of reality. First it must form (or find) a context for its contemplated actions. Then within this context, it selects a particular value for the quantum attribute evoked by that context.

Quantum Inseparability

Einstein's third New Physics peeve was quantum inseparability, probably the greatest surprise to emerge from the quantum worldview. The key to this surprise is the word *local*. All interactions both in classical physics and in quantum theory are explicitly local.

Locality means that when a body at location A acts on a second body at location B, the interaction must traverse the intervening distance. Furthermore, the velocity of this interaction must be no greater than the speed of light. The conventional way of imposing the locality requirement is via the notion of a field. To interact with body B, body A employs a go-between called a force field. Body A causes changes in this field that are propagated at or below light speed to body B. An example of a force field is the gravitational field between the sun and its planets.

The opposite of a local interaction would be a force that

traveled instantly from location A to location B without traversing the intermediate space. The only place that nonlocal forces played a role (before quantum theory) was in voodoo, whose practitioners believe that an action on a person's separated parts—his hair or fingernail clippings—can affect the whole man. Whatever voodoo witch doctors might think, such instantaneous leap-frogging effects have been universally rejected by all physicists from Galileo Galilei to Murray Gell-Mann. In accordance with this belief, the forces of quantum theory were designed from the start to be explicitly local, but the possibility waves that represent the particles possess a certain intrinsic "wholeness" that, in the mathematics at least, ties these waves together with unpleasant (to a physicist) nonlocal, voodoolike connections.

When two quantum entities, A and B, briefly interact (via conventional local forces) then move apart beyond the range of the initial interaction, quantum theory does not describe them as separate objects, but continues to regard them as a single entity. If one takes seriously this feature, called *quantum inseparability*, then all objects that have once interacted are in some sense still connected.

Unlike local fields such as gravity or electromagnetism, this lingering quantum connection is not mediated by fields of force, but simply jumps from A to B without ever being in between. Like a voodoo love charm, particle A is in touch with particle B because A's wave has kept a part of B's wave—its phase—in its possession. Because nothing really crosses the intervening space, no amount of interposed matter can shield the quantum connection. Since this nonlocal connection does not actually stretch across space, it does not diminish with distance. It is as potent at a million miles as at a millimeter. Just as a nonlocal connection takes up no space, so likewise it takes up no time. A nonlocal connection leaps between A and B immediately, faster than light. For some observers, as a consequence of Einsteinian relativity, this instantaneous connection appears to go backward in time, a performance peculiar by any standards.

Quantum inseparability, with its unsavory nonlocal connections, undoubtedly exists mathematically in the quantum
possibility wave formalism. But do these connections actually
exist in the real world?

Actual calculations show that even though quantum theory is connected nonlocally inside its mathematics, these connections never get out to the level of quantum predictions—
the only aspect of quantum theory that can be put to direct
test. These calculations predict that any measurable quantum
influence must travel at the speed of light or less. Thus, despite its inner nonlocality, quantum theory does not predict a
single nonlocal effect—as Philippe Eberhard, of the University
of California at Berkeley, first showed. In line with quantum
theory's perfect predictive success, no nonlocal connections
have ever been observed, either in the wild or in the laboratory. The perfect locality of all quantum measurements suggests that nonlocal connections are a theoretical artifact with
no more reality than the dotted lines that outline the constel-
lations on star maps.

Physicists continued to believe that these nonlocal connections were fictitious until the remarkable discovery of John
Stewart Bell. Bell showed in 1964 how the real nature of the
quantum connection could be put to experimental test. If the
experiments went one way, then the world must really be nonlocal; if they went the other, then one could continue to believe
that the mathematical connections were spurious. In 1970,
John Clauser and Stuart Freedman carried out Bell's experiment in Berkeley, confirming quantum inseparability. More sophisticated experiments by Alain Aspect in Paris increased
our confidence that the quantum world is really tied together
by nonlocal influences.

The Bell experiments involve two photons emitted back-
to-back from a common source. These two photons are created
in identical polarization states, described mathematically as
"phase-entangled possibility waves." One can visualize this
experiment by imagining two identical Schrödinger cats traveling at the speed of light in opposite directions, the "phase"

of one cat's wave inextricably entangled with the phase of her twin sister. The consequence of this phase entanglement is that when I choose a context for the measurement of cat A, either her mortality attribute or the X attribute, this choice of context instantly affects the outcome of the measurement of cat B. The upshot of the Bell-Clauser experiment is that any model of the world that does not contain nonlocal influences between the measuring device at A and the real condition of distant cat B must necessarily be an incorrect model.

Thus, despite physicists' traditional rejection of nonlocal interactions, despite the fact that all known forces are incontestably local, despite Einstein's prohibition against superluminal connections, and despite the fact that no experiment has ever shown a single case of unmediated faster-than-light communication, the Bell-Clauser experiment proves—although indirectly—that the quantum world accomplishes its tasks via real nonlocal connections. Bell's result requires that the world be filled with innumerable nonlocal influences in order to work as it does.

However, these Bell-mandated nonlocal connections are subtle. Although they travel faster than light, they cannot be used for signaling because they act on the level of individual quantum events, not at the level of patterns of quantum events. But the quantum events appear (to humans) to occur at random, the very opposite of a signal. Thus nature can use these superluminal connections for her own purposes, to knit the universe more closely together than was possible with local Newtonian force fields, but humans cannot decode these superluminal communications, which seem to us to be encrypted in an inscrutable random code to which only nature holds the key.

The significance of quantum inseparability for models of mind is twofold. First, the peculiar variety of wholeness possible for quantum systems may offer a possible mechanism for achieving the unity of experience observed in so many (human) minds. Second, the notion that mind operates by influencing the occurrence of otherwise random events gives rise to the

possibility that mind can influence distant matter in a decidedly nonlocal manner.

In the next chapters, we examine certain experimental
efforts to connect the inner life of the mind (both human and
otherwise) with quantum randomness, thinglessness, and inseparability.

quantum randomness: essence noise or subatomic spirit gate?

Uncertainty is the very essence of romance.
—OSCAR WILDE

Chance favors the prepared mind.
—LOUIS PASTEUR

Scene: Rudi's Artificial Awareness Lab.

NICK: So you've spent a lot of time meditating, Rudi?

RUDI: Meditation? You could call it that, Nick. For most of my adult life I've used my own mind as a laboratory. I've chased after gurus, sat for days staring at walls, twisted my body into knots, exchanged "chi" in a dozen forms of martial art, jacked my brain into exotic computer chips, and taken every drug I could get my hands on. I've also done a lot of reading. Have you seen my library, Nick?

NICK: What do you think you've learned, Rudi, from your years of self-exploration?

RUDI: For starts, I've experienced firsthand the two main illusions that mind science has to offer: materialism and idealism, the "Isaac Newton" trip and the "Bishop Berkeley" trip. At one time or another, I've known each one of them to be true.

In states of ordinary awareness, it's easy to fall for the Newton trip, to convince yourself that matter is all that there is, and that mind is just a mechanism made of meat, fragile and unimportant, a lucky biological accident in an otherwise heartless mechanical universe. The main reason that the Newtonian illusion is so persuasive is a matter of numbers, I think. So much of what our bodies do is done without our being aware of it. Unconscious processes in the brain overwhelm the conscious ones by a ratio of more than a trillion to one. In bodies like these, it's no wonder that mind seems insignificant compared to matter.

NICK: But the materialist viewpoint has produced tangible results. How can you call it an illusion? Our science is based on the fact that we can do experiments on matter without regard to that matter's inner experience or the inner experience of the scientist. That's what we mean by objectivity in science: its truths are universal, the same for everyone, free of the taint of personal subjectivity. Doesn't the immense success of the scientific method validate to some extent the existence of a material world, a world outside us, independent of our dreams and desires? Could there be any science at all if reality depended strongly on the scientist's state of mind? How can anyone these days believe in Berkelean idealism except as some sort of playful exercise of the imagination?

RUDI: I could feed you any of a dozen "mind medicines," or hook you up to a "trip chip," and your confidence in the Newtonian worldview would collapse like a soap bubble. In certain altered states, it becomes self-evident to me that everything is made of mind, that matter consists of fleeting vibratory patterns in some vast field of consciousness. Furthermore, my conviction that everything is made of consciousness is validated in the most scientific way possible, by direct experience: in this state I feel with my own being how the world arises moment by moment out of a mental substrate. Moreover, the clarity and intensity of this privileged access to the inner life of the universe completely overshadow all my previous observations and theories. External sensory knowledge seems pale and secondhand compared to this direct experience of the world's inner being. In my state of Berkelean rapture, I know for a fact, with unshakable certainty, that all perception is extrasensory; that the universe is wholly personal, made up of mental entities like me; and that this universal mentality is indestructible: it can never die.

Of course, when my brain returns to normal, the Newtonian illusion takes over again and relegates the Berkelean vision to the status of a philosophical hallucination.

NICK: But the Newtonian model works: look at all the reliable technology that arises from the materialist hypothesis. If the universe is really mental, why haven't mediators and cognitive psychologists managed to move the world closer to their heart's desire?

RUDI: I believe that both visions are illusions, Nick, part of a larger truth. The world of mind needs matter as a relatively stable medium in which to express itself, and the material world needs mind to make its existence "meaningful." As for mind-created reality, it's obvious to me that our technological accomplishments result from the interaction of a particular kind of mentality with matter. Our culture is not entirely material but a co-creation of mind and matter. I imagine it's the same in the nonhuman world as well.

NICK: But if mind is really present everywhere, why aren't robots conscious?

RUDI: I don't know the answer to that, Nick. That's why I continue to experiment with artificial awareness.

If the universe really does consist of interpenetrating mental and physical worlds, then we might expect that the laws of each world are conditioned by those of the other. We know that human mental life is strongly affected by the material condition of the brain. Are physical acts likewise shaped by the inner lives of invisible beings? In particular are the laws of quantum theory a public reflection of innumerable private experiences? The notion that behind every physical process lies an invisible mental experience might be called the hypothesis of "quantum animism." In this view, a system's possibility wave represents the range of action—or realm of possibility—open to the conscious being inside that system. Every quantum wave is the potential home of some form of consciousness, and vice versa: where there's a will, there's a wave.

Classical and Quantum Systems

If only quantum systems are conscious, then what constitutes a quantum system? A quantum system may be distinguished from its classical cousins by the number of possible courses of action that are in fact open to that system. If there is only one possible outcome, as, for instance, in a deterministic computer program, then the system is a classical Newtonian one. If, however, there is more than one possible outcome, the system is quantum.

Certain systems, such as a pair of dice tossed on the table, may be so complicated that we cannot predict the outcome, but we know that this outcome is theoretically fixed by the initial conditions, so that once these are specified, only one outcome is possible. Imagine a "control space" labeled by all possible initial-condition variables. Each point in this space represents a slightly different way of throwing the dice. We can imagine the control space painted with 36 different colors corresponding to the 6×6 different possible dice outcomes.

Certain solid-colored regions of this space may correspond to the outcome 4 + 3: if the dice are thrown anywhere in this region, they will always turn up 7.

The new field of chaos mathematics focuses attention on regions of the control space where different colors are braided together in an intricate filigree, threads of 4 + 3 tangled up with threads of 4 + 2 and 5 + 3, for instance. In these "regions of chaos," a slight change in the initial conditions leads to a drastic change in the final outcome. However, even here, in the region of chaos, a single initial condition always leads to a unique outcome.

Chaos mathematicians assume that the dice are governed by the laws of Newtonian physics, which guarantee a single deterministic output for any well-defined input. But, in reality, the dice are governed by quantum rules that introduce a tiny uncertainty into the initial conditions as well as the trajectories that follow from these conditions. In the solid-colored region of the control space, the quantum uncertainty has no effect: here Isaac Newton rules. However, in the chaos region of the dice control space, wherever the magnitude of quantum uncertainty is greater than the "thread size" of the chaotic filigree, then the dice system certainly follows quantum rules. Dice tossed in this region do not have a unique outcome. Here, for instance, the outcomes 4 + 3, 4 + 2, 5 + 3, and several others are all simultaneously possible. The control space can be conveniently divided into three regions: a classical region characterized by large volumes with the same color, a classical-chaos region characterized by braided filigree whose thread size is larger than the quantum uncertainty, and a quantum-chaos region where the quantum uncertainty is larger than the thread size. In the classical part of the control space, a single die toss has only one possible outcome; in the quantum-chaos region, a single die toss truly has many possible outcomes.

In general, the range of quantum possibilities is set by the Heisenberg uncertainty principle, which is scaled by Planck's constant of action, an exceedingly small number compared to the actions of everyday events. An eyeblink, for instance, represents an amount of action of about 1 erg-second. On this scale, Planck's constant is 10^{-27} times smaller: there's a billion times a billion times another billion Planck units of possibility in the blink of a human eye. For ordinary events, the leeway afforded by Planck's constant of action is insignificant. For all practical purposes, most ordinary events have only one possible outcome. The smallness of Planck's constant explains why Newtonian physics worked so well for such a long time.

Quantum possibilities contribute in a significant way to the gross motion of matter in only two situations: when the energy of the interaction is small, as in the case of atoms or molecules, and in "quantum razor" situations where the system's possibility wave is split into two or more disjoint parts, each having drastically different experimental consequences, such as Dr. Schrödinger's legendary live/dead cat.

Quantum systems are characterized by randomness, thinglessness, and inseparability—all absent in a Newtonian system. To the quantum animist, quantum randomness is not random at all but represents the opportunity for the exertion of free choice by some mindful being. Quantum thinglessness describes the peculiar status of an unobserved quantum system: such systems consist of context-dependent possibilities, not fixed actualities. The profoundly ambiguous state of an isolated quantum system must correspond to the way in which conscious beings, including us, perceive themselves and the external world. The objective thinglessness of quantum systems implies the subjective thinglessness of elemental minds. Finally, quantum inseparability provides a new way of tying together distant systems, a way not available to classical systems. We should expect the quantum connection to shed new light on the observed unity of single minds and the alleged distant communion of mind with mind outside conventional sensory channels.

Despite the fact that many scientists—for instance, Hungarian polymath John von Neumann, Nobel laureate Eugene Wigner, Berkeley physicist Henry Stapp, Princeton physicist Freeman Dyson, and Oxford mathematician Roger Penrose—have expressed the belief that a deep connection exists between mind and quantum matter, surprisingly little experimental work has been carried out to verify or disprove the quantum animism hypothesis. Most physicists continue to believe that quantum theory is consistent with a purely materialistic view of the world and that there is no necessity or advantage to bring consciousness into physics through the back door provided by the Heisenberg uncertainty principle. Heisenberg himself, for instance, saw no relationship whatsoever between consciousness and modern physics. Because of this widespread indifference to the quantum animism hypothesis, the few experiments actually carried out to test quantum mind/matter hypotheses have been tongue-in-cheek, a playful extension of the scientific method. Of course, if any of these playful experiments were actually successful, the game would immediately turn serious, the players be dubbed "bold pioneers," and physics suddenly be catapulted in a radically new direction.

Quantum Randomness: The Denver Experiment

A beam of visible light was the first system discovered to possess quantum properties. Experiments carried out over a period of more than a hundred years had clearly shown light to be a wave phenomenon. However, at the beginning of this century, Einstein showed that the photoelectric experiment—the ejection by light of electrons from a metal surface—could be concisely explained by considering the light to be a beam of particles now called *photons*. Thus light, when it is moving (unobserved) from place to place, acts as a wave; when it is observed—by interacting with the electrons in a metal plate, for instance—it behaves as a particle.

All quantum entities act this way: as a wave when unobserved, as a particle when viewed. Every quantum phenomenon—and all phenomena without exception are believed to be ultimately of quantal origin—has both a wave and a particle aspect, the particle aspect labeled usually by the suffix -*on*, the Greek word for "entity" or "thing." Thus the particle aspect of sound, light, electricity, and magnetism is called "phonon," "photon," "electron," and "magnon." The hundred-odd fundamental "particles" ("wave/particles" would be more accurate) discovered with the aid of high-energy accelerators include tauons, pions, bosons, muons, and gluons. A few particles, such as the quarks, have inconsistently escaped this ontic nomenclature, but, in general, when you read about some new kind of "-on" in physics (graviton, soliton, proton, neutron, and so on), you can be almost certain that it refers to the particle aspect of some quantum wave.

The unmistakable sign of a wave, quantum or otherwise, is that when you try to send it through a narrow slit whose width is of the same order of magnitude as its wavelength, the wave emerging on the other side will form a *diffraction pattern*: the wave fans out from the slit and exhibits diffraction maxima and minima whose precise location and intensity depend only on the wave's wavelength and the width of the slit. Thus a diffraction pattern both reveals the presence of a wave and allows you to measure its wavelength.

To observe the diffraction pattern of a light wave, for instance, one must place some sort of light detector behind the screen. Any light detector—photographic film, human eye, or TV camera, for instance—if it possesses fine enough resolution, will always reveal the light arriving in the form of tiny packets of energy, precisely localized in time and space. The diffraction pattern shows light to be a spread-out wave; the detection of tiny impulses of energy shows it to be a concentrated particle. This twofold style of existence displayed by light in the diffraction experiment is characteristic of all experiments with quantum systems.

In 1970, a group of students at the University of Colorado in Denver decided to use the diffraction experiment to test whether photons are conscious. Experimentally the arrival of a photon at the diffraction detector appears quite random. It is completely unpredictable, for example, whether a photon will be detected in the right or in the left half of the diffraction pattern. On the theoretical side, quantum theory precisely predicts the shape of the diffraction pattern, which is due to a large number of photons, but regards the individual events that make up the pattern as utterly random. One possible consequence of the quantum animism hypothesis might be that each photon is endowed with a mind of its own that selects in some way the direction in which it will bend when it goes through the slit.

In their paper entitled "Photon Consciousness: Fact or Fancy?" the students argued that once a particular photon had made the choice to bend in a particular direction, then, when confronted with a second narrow slit, it would persist in its choice and show a tendency to favor the previous diffraction direction.

This hypothetical tendency of conscious photons to persist in their choices could be tested by sending a portion of those photons that had diffracted to the right through a second slit. If the photon consciousness hypothesis is valid, the second diffraction pattern will not be symmetric but will show an excess of photons on the right-hand side. According to quantum theory, photons possess no memory, so the second diffraction pattern should be as symmetric as the first. Although simple to do, a double-diffraction experiment of this kind had apparently never been performed.

The students set up a pair of slits, the first slit illuminated by a laser beam, the second slit illuminated by various segments of the first slit's diffraction pattern. In all cases the results agreed with standard quantum theory: the second pattern was completely symmetrical. The photon consciousness hypothesis, at least in this form, was refuted. If photons possess consciousness and use it to select their diffraction direction, they do not appear to possess a memory of their first choice that influences their subsequent behavior. The negative results of this imaginative experiment no doubt encouraged the students to pursue more conventional kinds of research; no further work on photon consciousness was ever reported.

Quantum Randomness: Schmidt Machines and Their Kin

Although the main concern of *Elemental Mind* is ordinary awareness, the everyday inner life of humans and other conscious beings, much can be learned about awareness from rare and unusual states of consciousness. Foremost among the paranormal powers of mind is psychokinesis—the alleged ability of certain minds to reach out and affect distant material systems without the mediation of physical forces. Psychokinesis (PK) is important to theories of consciousness because the very existence of such an ability would immediately refute en masse all simple mechanical models of consciousness, eliminating from serious consideration, for example, computer-based models of mind such as Marvin Minsky's.

Spontaneous cases of psychokinesis of the poltergeist (German for "noisy ghost") variety have been investigated for decades. In these situations, a single person, usually an adolescent, is the focus of the occurrence of loud noises, moving or flying about of heavy objects, and unexplained disturbances of electrical equipment. The Rosenheim poltergeist, for instance, which took place in a Bavarian law office in 1967, was centered around Annemarie, a 19-year-old employee of the firm. The effects included flickering of light bulbs, which also unscrewed themselves from their sockets and exploded, as well as movement of heavy filing cabinets and strange percussive noises. The Rosenheim phenomenon was studied by investigators from the electric company, two physicists, and a professional psychic investigator. Some of the phenomena were recorded on videotape, but no normal explanation was ever discovered.

Poltergeist cases suggest the existence of a psychokinetic power, but their sporadic occurrence and short duration make them difficult to study scientifically. To gather more data on this elusive phenomenon, laboratory tests on psychokinesis have been carried out, beginning with the work of J. B. Rhine at Duke in the 1930s, who reported some success in observing a psychokinetic effect on dice and other physical systems.

Because quantum systems are considered completely random, utterly immune to any known physical influence, one might imagine that they would make ideal subjects for psychokinetic testing. If photons are not themselves conscious, as the Denver experiment suggests, perhaps they are open to outside mental control. Physicist Helmut Schmidt, formerly at Boeing Corporation in Seattle, now at Mind Science Foundation in Austin, has built several quantum-random display devices to test the hypothesis that human intention can influence quantum-random events.

One variety of "Schmidt machine" is a box with an on/off switch and a circle of a dozen lights. When the machine is running, one light is always lit, and then in a seemingly random manner, that light goes off and an adjacent light goes on. Which of the two adjacent lights (right or left) turns on is determined by the time of emission of a single quantum particle from a radioactive source. The average emission time of a radioactive element is fixed by quantum rules, but the emission time of a single particle is completely unpredictable. Running by itself the light seems to hop around the circle at random with equal preference for the clockwise and counterclockwise directions. Computer analysis of an unattended Schmidt machine shows no deviation from the "rules of randomness" established by statisticians.

To test for the presence of psychokinesis, a human subject watches the lights and tries to "will" the light to rotate in a selected direction. Most people are unsuccessful at this task, but certain subjects at certain times have been able to move the lights in a particular direction in such a manner that the results exceed what would have been expected by chance by the odds of several thousand to one.

The Princeton Experiments

Recently, Robert Jahn and Brenda Dunne at Princeton have carried out similar random-number generator (RNG) tests with trials consisting of millions of events. They have achieved results significantly above chance both for selected subjects and for the average scores of all subjects combined. Although their work seems to validate the existence of a psychokinetic effect, it does not necessarily support a simple quantum consciousness hypothesis. In addition to RNGs based on quantum randomness, Jahn and Dunne used as targets systems with nonquantum sources of randomness, including a kind of pinball machine filled with polystyrene balls, as well as runs of completely deterministic random numbers generated by computer. The size and quality of the PK effect did not seem to be device-dependent.

Critics of psychokinesis are fond of pointing out that objective, well-funded laboratories for testing the PK hypothesis already exist in Reno, Las Vegas, and other casino cities around the world. Both the dice and roulette tables represent attractive targets for those who believe they possess the ability to move matter with their mind. Can gambling houses be exploited not only to verify psychokinetic abilities but to provide a source of funding—random money generators, so to speak—for further research into parapsychological powers?

Let's look at the facts. For any public gambling game, the odds are always in favor of the house. Many people win but more people lose: the house edge is what keeps casinos in business. To test your PK abilities in a gambling house, you will want to choose a game for which the house advantage is the least. Here are three possibilities. Betting red or black on an American roulette wheel (which has both a green 0 and a green 00): house odds = 2.56 percent. Betting red or black on a European roulette wheel (which has only one green 0): house odds = 1.35 percent. The game of craps, in which the player tries to make several kinds of winning dice combinations before throwing a losing combination: house odds = 1.41 percent.

These numbers represent the amount of psychokinetic ability that you would have to muster in order to convert a loser's game into a winning proposition: it would take a "psychic force" of roughly 2 percent, that is, out of every 100 turns of the wheel, 2 events on the average change from their chance value to that of the "mind's desire." I will call the figure 2 percent the "Reno minimum" for psychic ability to be able to turn a profit at a gambling house. What is the magnitude of PK measured in the laboratory? Sad to say, it is far short of the Reno minimum.

Most people in Jahn and Dunne's study scored in the intended direction but very close to chance expectation. A few PK stars scored as high as four standard deviations in the intended direction, achieving odds against chance of close to a million to one. However, these extraordinary scores were achieved by maintaining small percentages consistently over a large number of events. The best PK performance to date in the Princeton experiments amounted to deviations from chance expectations of at most 0.1 percent, or one part in a thousand, approximately twenty times smaller than the Reno minimum. Most people did considerably worse than this.

Psychokinesis has important implications not only for consciousness studies but also for physics. If the mind can indeed exert a force on distant matter, then current physics is demonstrably incomplete since it recognizes no mind-based forces whatsoever. To assess the scientific status of psychokinesis, Dean Radin and Roger Nelson at Princeton undertook a data base search of all reported PK studies and subjected them to a statistical meta-analysis designed to quantify the presence or absence of a PK effect in this mass of data produced by many independent researchers using many different systems and procedures.

Radin and Nelson concluded that the data robustly suggest the existence of a PK effect whose average magnitude is of the order of 0.02 percent (100 times smaller than the Reno minimum) and that these data cannot be explained away by fraud, the "file drawer effect" (unreported negative experiments), or poor experimental design. They suggest that physicists take seriously—their paper was published in a prominent physics journal—the possibility that mind can directly affect the motion of distant matter. The smallness of the PK effect should not minimize its importance. One of the most fundamental and still mysterious discoveries of the twentieth century is the fact that the decay of K-zero mesons violates "time-reversal invariance"; this reaction (and no other) possesses a built-in arrow of time, to a small but undeniable degree—about 0.2 percent—of the same order of magnitude as the scores of the best Princeton PK subjects.

Can PK Be Amplified by Averaging Over Many Trials?

One of the main experimental problems of PK research (and parapsychological research in general) is the sporadic nature of the results. Experiments do sometimes yield odds against chance of millions to one, but such performances are difficult to repeat. In the jargon of the information theorist, we seem to be dealing here with a noisy communication channel. However, the founder of information theory, Claude Shannon, showed how one could reliably send messages along any channel no matter how noisy: one simply repeats the message again and again. Over the long run the noise averages to zero, while the signal steadily increases. With enough repetition, any signal can be reliably sent through even the noisiest channel.

In a random binary process consisting of N events, the standard deviation, or "noise," is proportional to $\sqrt{N}$. This means that if you toss 100 pennies, the result will be 50 heads $\pm$ 10 more than two-thirds of the time. If you toss 10,000 pennies the result will be 5000 heads $\pm$ 100.

In the first case the "noise" is 10/50 = 20 percent; in the second case it is 100/5000 = 2 percent. As the number of tosses increases so does the noise, but the noise increases more slowly than the number of tosses. As the number of tosses gets larger, the ratio of noise to tosses gets smaller so that the relative spread in values around the average decreases, a particular example of what statisticians call the *law of large numbers*—the tendency of most statistical processes to converge relentlessly to their average values as the number of events increases.

From this example, one can see that if one tosses enough coins, the relative "noise" can be made as small as one wishes. For instance, after 10 billion tosses, the noise has been reduced to $100{,}000/5{,}000{,}000{,}000 = 0.002$ percent or ten times smaller than the average PK effect as calculated by Radin and Nelson.

This simple PK-as-signal model assumes that the PK effect behaves as a conventional "signal," that is, that the mind pushes against the chance distribution with a constant pressure, expressed as a certain percentage K of the total number of events N. In other words, in the PK signal model we tacitly assume that the PK effect X (average number of mind-modified events) can be expressed as $X = KN$ where K is of the order of a few hundredths of 1 percent for the average PK subject.

Attempts to amplify parapsychological effects via Shannon theory by increasing the number of psychics or the number of trials, by polling, or by covert repetition of the same experiment have been generally disappointing. If a successful method for extracting the "PK force" from noise were ever discovered, it would immediately be utilized by psychic researchers to decrease the noise in their experiments. Rough examination of the cumulative data, thoughtfully provided in pictorial form by Jahn and Dunne, seems to indicate that they are consistent with the notion that the PK effect X (number of mind-modified events) is not proportional to the number of events N, but is at most proportional to $\sqrt{N}$. In other words, $X = K\sqrt{N}$. This hypothesis amounts to the assumption that the "PK force" is proportional not to the number of events, but to the "noise" present in the system. One unpleasant feature of this psychic noise hypothesis is that as the number of events N increases, the relative strength X/N of the PK effect actually decreases, inexorably smothered, like the noise to which it is proportional, by the statistician's law of large numbers. Now that Radin and Nelson have more firmly established the actual existence of a PK effect, an important goal of Schmidt machine research and its successors should be the determination of how the PK effect X actually does vary with the number of elemental events N. Does the "relative push of a wish" X/N against a parade of random numbers persist, or does it fade away as the number of such numbers increases?

Seven years of experimental work at Princeton produced an overall PK effect with all subjects taken together that was so far from random expectation that the odds that such an effect will occur by chance are almost a million to one. This means that, if experiments of this sort had been continually carried out since the Stone Age, no more than one result this far from the average would have occurred by accident. The Jahn and Dunne experiment results are certainly statistically significant, more so in fact than some physics experiments. But can these extraordinary results be repeated in other laboratories? Because of the Princeton experiment's consequences for a theory of mind—if these results are true, mechanistic mind models are highly unlikely—it is important to verify Jahn and Dunne's remarkable claim that the minds of ordinary people can influence matter, to a small but undeniable degree, without the mediation of known forces.

Quantum Randomness: The Metaphase Typewriter

In 1963, Jane Roberts, an obscure writer in Elmira, New York, became the mouthpiece for a discarnate entity that called itself "Seth." Seth claimed to dwell in a world of "probable selves," the world out of which present, past, and future incarnations of human personalities arise. To a physicist, Seth's world of "probable selves" is reminiscent of the unobserved quantum world of "possible attributes," which is by definition unobservable by conventional measurement devices. The Seth personality, whatever its true nature, subsequently became the subject and the author of more than a dozen books on the nature of personal reality.

In the 1970s, Seth-type phenomena increased as many individuals became channels for discarnate personalities. Some of these entities claimed to come from other (inevitably higher) dimensions and some from the stars. None seemed to come from the same realm as Seth, so his messages ended with the death of Jane Roberts in 1984.

The increase in the number of active channels led Seth and Jane's editor, Tam Mossman at Prentice-Hall, to start a scientific journal on channeling called *Metapsychology: The Journal of Discarnate Intelligence*, a kind of "Nonphysical Review" for public exploration of an area of inquiry left completely untouched by *Physical Review*, the major American physics journal.

At the beginning of this century, a similar rash of discarnate communicators appeared and were studied by Harvard psychologist William James and members of various psychical research societies. These discarnates were not from other stars or dimensions but claimed to be the souls of humans who had recently passed away. For mind scientists, one of the most exciting kinds of experiment conducted by these early researchers was the "cross-correspondence" phenomenon, in which the same discarnate entity spoke through two or more different mediums. Needless to say, the reception of the same or similar information by two physically separated mediums would constitute highly evidential support for the discarnate's claim that it was acting from a dimension that is independent of ordinary space and time. As far as I know, none of the modern discarnates has ever attempted to speak through more than one human channel.

In the early 1970s, members of the Consciousness Theory Group (centered at that time in Berkeley, California) were fascinated by the growing discarnate intelligence phenomenon. One of our concerns was the ethical propriety of one entity's occupation of an already spoken-for body, even with the first occupant's (conscious) permission. We are complicated beings, after all, and permission given with one part of the mind may not be echoed by other parts—parts that may experience discarnate occupation as a rude psychic violation of privacy. A few of us at CTG decided that, certainly for ethical reasons and perhaps for scientific reasons as well, it would be better if discarnates could enter this plane of existence through a vehicle that was not currently being occupied by a sentient being. The goal of the "metaphase typewriter" project was to make available a mechanical or electronic communication channel for discarnate entities, a channel that was initially empty of sentient experience.

The design of so-called metaphase devices was inspired by physicist Heinz Pagels's conception of quantum theory as "the language of nature." We simply took Pagels's metaphor one step beyond what the author of the celebrated *Cosmic Code* had originally intended.

If one examines the text in this book, for instance, one will find that spaces between words occur about 17 percent of the time, the letter *e* occurs about 11 percent of the time, and *t* makes an appearance 8 percent of the time, making up a distribution of letter frequencies that is surprisingly stable and independent of the content that these words express. A person analyzing this text with the tools of a statistician will end up with tables of statistical data that in some sense completely describe the way letters are used in this book. But no matter how exhaustive the statistician's letter counts, they entirely fail to grasp the book's main purpose: the coding of meaningful information in nonstatistical ways.

The quantum animism hypothesis assumes that every quantum system is, or could be, alive, that is, possessed by some (currently invisible) inner experience. The behavior of such systems is described by quantum theory in a statistical manner, completely codified in the Schrödinger wave function, knowledge of which allows the physicist to calculate the probability of the result of any measurement one might choose to perform on the system.

Suppose we imagine that the quantum statistics are a kind

of language statistics for the conversational behavior of various subquantum sentient entities. The physical world would be, in this view, spoken into being by a vast interconnected community of invisible voices. The import of these quantal voices lies not in the statistical distribution studied by physicists any more than the import of this book lies in the number of *e*'s and *t*'s it contains. Rather the inner meaning of a quantum system resides in the individual quantum event: the very thing that happens, not the after-the-fact statistics of many such events. These individual events—if this hypothesis is valid—are not random at all but represent the languagelike behavior of numerous sentient beings.

To test this quantum-events-as-language hypothesis, I built, along with graphics engineer Dick Shoup, then at PARC Xerox in Palo Alto, California, a mechanical device that translated certain elementary quantum events into individual linguistic units of the English language. The metaphase typewriter consisted of a quantum-random system, an interface that reinterprets the random events according to some preconceived code, and an output device meaningful to human beings. The first proposal for a metaphase type machine was that of Alfredo Gomes, a Brazilian physiologist, who suggested connecting the random events that make up an electron's diffraction pattern to a piano, a sort of quantum jukebox that could eavesdrop on subatomic music festivals. To my knowledge, the quantum piano was never built. The metaphase typewriter, on the other hand, achieved first contact with the quantum world on January 10, 1974.

At the heart of the metaphase typewriter sits a radioactive source, thallium 204, with a half-life of about 4 years, that decays into lead 204 by emission of a beta ray (physicists' slang for a very fast electron). The beta ray is detected by a Geiger counter, producing, over time, a series of electronic pulses—one for each beta ray, whose individual time of occurrence is highly unpredictable but whose statistical properties are well known. Since, for each run, we always adjusted the detector to obtain a counting rate of 60 counts per second, the average pulse separation was always about 17 milliseconds.

Around this average value, the time intervals between pulses fluctuate wildly, but some intervals (short ones) are more probable than others (long ones), following a statistical law called the Poisson distribution—one of the many "rules of randomness" that govern the average behavior of random processes such as coin flipping and dice games. We used the Poisson distribution to produce a translation code that would generate English text whose second-order letter distribution was identical to that of ordinary written English. Suppose a t has already been printed; then h is very probable, u less likely. We designed the "language filter" in such a way that if a very probable Geiger pulse interval occurred, then an h would be printed after the t, and improbable Geiger events printed less probable letters. Our source for the English language statistics was unclassified document #S-209,179 of the National Security Agency, who presumably compiled these numbers for more serious purposes.

The metaphase typewriter was designed to support three kinds of output devices: a text generator, which was an actual typewriter; a speech synthesizer to produce quantum-based vocalizations; and a number of graphic displays. Both the typewriter and the speech synthesizer were actually built and operated in various "high-energy" psychic environments with uniformly disappointing results. When the metaphase typewriter was turned on it produced text at a rate limited by the typewriter's mechanical action. Because of the built-in second-order statistics the text somewhat resembles English, or a half-solved cryptogram. Here is a small sample of metaphase text:

WIRN OF ACERIONINE SE IND BE B WHAD ATHE ORO-
VESSOUNDRO MAT PIND ASPAS HESUN UR D T CORE
G LVIDESPANOUMO BIMARNAGLES HSTEAF NNAN A
AITHIDIF PUTAMSUBENES T QUALOA ASELOTNULARE

INE T THAPE ALLIGACAZOF WANE HT F A T G R ATHE
FOVA WHISERDEM INOT ACRYRYIVESSTHENEMBOFO
OR W WO WOMAD FORDISP AS HE WHA CO T T PLE F
T OWRUS INIAIDITHE COR NITAL PIS D BEANSTO AR-
ERS THESITIVENOVERLASESTEWONM IST MIGHIPOF A
DUNKISHENT ISEAD RIENDUBE THERROIN

For our experimental design, we conceptualized the typewriter in two ways: (1) as a PK device similar to a Schmidt
machine, but with a somewhat more interesting output
display—English pseudotext or spoken syllables; (2) as a mechanical medium, a newborn, unoccupied psychic channel ripe
for takeover by some wandering discarnate without the ethical
restrictions that might accompany occupation of a human host
mind. The metaphase typewriter might provide, we imagined,
the potential for the creation of a thought-operated word processor (every writer's dream) or, in a more flamboyant mode,
an "open mike to the Void"—an English language channel
for discarnate intelligences from anywhere in the cosmos:
between-life Buddhist Bardos, the Islamic seventh heaven, or
the physicist's eighth dimension.

Metaphase Psychokinesis

The metaphase typewriter as PK text generator was tested
by two notable psychics, Englishman Michael Manning and
San Franciscan Alan Vaughn. Manning disliked the speech
synthesizer output, which writer Robert Anton Wilson described as sounding "like a Hungarian reading *Finnegans
Wake*." He concentrated instead on trying to influence the
quantum text generator in some remarkable but unspecified
way. During Manning's afternoon session with the metaphase
machine, no untoward behavior was observed.

Our subsequent session with Alan Vaughn was less infor-
mal. Before Alan's arrival we prepared several 3 × 5 index
cards on each of which was printed a single target word:

ITALY, IGLOO, KNIFE, for instance. Alan was given one of these cards and left alone in the typewriter room to attempt mentally to impress that word on the "DADAstream" flowing out of the quantum possibility waves of radioactive thallium. None of the words appeared in the text during the test periods, although a few unusual events occurred before and after what has come to be called "the ITALY experiment."

After the test, before we had turned off the machine, the phrase ITAL Y appeared, as well as the phrase BY JUNG. "I wonder where that came from," Alan laughed. "Maybe this could explain it," said a young lab assistant as she pulled a paperback edition of Jung's works from the pocket of her white lab coat—a curious and amusing event but not remarkable enough in my eyes to be considered an unambiguous psychic hit.

To prepare the index cards for the ITALY experiment, we decided to let the typewriter itself select the target words. We turned the machine on, copied down the first several "words" that it typed (ITHE, KNGANGHTH, WEDIS, FFIN, FINT, IG). Then we looked up these "words" in a big 2300-page unabridged dictionary in the laboratory library. Since none of the "words" was a real English word, we chose the nearest suitable dictionary entry to be our target word. When we got to the library, however, we found the dictionary already open to the very page indicated by the first metaphase "word" (ITHE), from which we generated the first target word ITALY. This curious coincidence—the outside world seemingly conspiring to ease our task—is suggestive of writer Arthur Koestler's "library angel," his name for similar literary coincidences that have helped him and others, against great odds, to locate obscure books relevant to his research interests. Whatever the nature of these "helpful coincidences," they in no way validate that version of the quantum animism hypothesis that we set out to test: the notion that quantum-random systems can be reliably influenced by human minds.

Metaphase Seances

For our experiments with the metaphase typewriter as mechanical medium, we set up the laboratory as a seance room and invited a particular personality recently deceased or known to be interested in after-death communication to take over the typewriter. My friend Bill Kautz, who has for many years been investigating ways in which psychics and scientists can become research partners, visited Jane Roberts, described our experiment to "Seth," and asked for his help: Would he or one of his discarnate associates be willing to participate in a scientific experiment to link the "probable selves" of his psychic world with the "probable quantum states" of a text-generating radiation source?

Seth replied that he was interested in people not machines and showed no further interest in the metaphase project. While the project was active we found no other discarnate consultants willing to cooperate in improving communications between the spirit and material worlds by quantum-mechanical means. An article published in *Psychic* magazine, describing the project and asking for volunteers (embodied or discarnate), elicited no replies.

Our most elaborate metaphase typewriter experiment took place on April 6, 1974, the 100th anniversary of the magician Harry Houdini's birth. Houdini was intensely interested in mediumistic communication and promised to send a message, if he could, from the "other side." For 10 years after his death, seances were held on Hallowe'en (Houdini's death day) with only one tangible result, a message generally regarded as fraudulent, via the medium Arthur Ford.

Before the Houdini seance, posters were widely distributed issuing the following challenge:

A Heisenberg-uncertain typewriter has been set up at an undisclosed Northern California research center. Its sensitive inner quantum mechanism appears to be free enough from every known physical law to permit takeover as a communication terminal by a sufficiently skillful discarnate entity. Metaphase type-writer is a presumptive open mike to the Void. Should you decide to accept this challenge, HARRY HOUDINI, and successfully impress your intentions upon the stream of random anagrams endlessly flow-ing from the teleprinter, you will be warmly wel-comed by our little band and most justly ranked among the great masters of escape.

Pictures of Houdini in various constraints were posted in the seance room: Houdini in chains, behind bars, in a strait-jacket, tied to a ladder. We doused the lights, held hands, med-itated, danced, sent out for pizza; some of the participants even took LSD. Children were running through the halls shouting for Houdini, while their parents chanted invocations to coax the magician's spirit to enter into the erratic quantum rhythms ticking away at the heart of the metaphase typewriter.

As with the ITALY experiment, the most unusual event occurred outside the formal experimental protocol. After the program was loaded and various technical problems solved, I pushed the reset button and the typewriter sprang into action. However, the typewriter's paper feed was jammed and lines of text were printing in a haphazard manner, at various angles across the skewed paper. The disorderly lines formed a roughly elliptical frame in which a single line of text was set. "ANINININFINITIME," it said—possibly meaning that in an infinite time we would certainly get a message from Hou-dini (and every other possible message too). This cryptic result also recalled the tale of the hundred monkeys sitting at word processors: eventually (after a time much longer than the age of the universe) the monkeys will succeed in producing all the works of Shakespeare.

Although the Houdini experiment failed to validate the quantum animism hypothesis, the ANINININFINITIME re-sult convinced me that the universe does possess a strange sense of humor. Although the metaphase project shed no light on the relationship, if any, between quantum randomness and the willed actions of conscious beings, perhaps this account of past experiments bearing on the quantum animism hypothesis will inspire more imaginative and successful experiments in the future.

quantum thinglessness: subatomic double-talk or wave logic of consciousness?

Prima Materia, if it is to be used for human purposes, must be "fixed" in a stable substance capable of being handled.

—HERMES TRISMEGISTUS

When a pickpocket looks at a saint, all he sees is pockets.

—BABA RAM DAS

Scene: Betsy's Bionic Boutique, a cross between a beauty parlor, an electronics warehouse, and an auto body shop. This is the place where robots, cyborgs (metal-flesh hybrids), and daring humans go to modify their bodies for decorative, occupational, sexual, recreational, and decline-to-state reasons.

BETSY: Oh, Claire, you're here at just the right time. Copies of the latest brain polyp styles from Philadelphia just came in the door. The optics are dazzling; at theta frequency your head just dissolves into a throbbing gold-green haze. I'm dying to see how it looks on you.

CLAIRE: Oh, Betsy, I don't know. Maybe it'll help me forget about my date with Rudi this weekend.

BETSY: Rudi's OK, Claire. Just a bit obsessive about giving

robots minds, that's all. I don't see what all the fuss is about. For 20 years I've gotten along perfectly fine without a mind. What's the big deal?

CLAIRE: Rudi says that robots are empty-headed sleepwalkers. He wants to wake me up; he wants to give me "inner experiences" like his own. That outlook tree machine he welded in last week didn't work, as far as I could tell. Now he wants to hook my brain to some quantum-random chip called an "Eccles gate."

BETSY: Quantum-random chip? Is that something like the "fuzzy logic" fad those Chinese robots started a few years back? I think I still have some of those Chinese chips in stock.

CLAIRE: No. Fuzzy logic is just more of the same old deterministic hardware. Quantum logic, on the other hand, is supposed to tap into the very structure of the atomic world. According to Rudi, a quantum brain drinks up waves from an ever-present background ocean of pure possibility, the ocean out of which comes everything, mind and matter alike. I wish I knew more about quantum mechanics. Can you dig up for me somewhere an advanced physics ROM? I think you're right, Betsy. That brain polyp does go nicely with my skin coloring. I'd like to try it on, the green-gold one, yes.

BETSY: Sit down over here, Claire, and let me open up your braincase. What good is consciousness anyway? Why would a robot ever want to have a mind? What can a human do that a properly programmed robot can't do better and faster?

CLAIRE: We seem to lack some sort of particularly human kind of internal processing. Nick claims that I detect things but don't perceive them; that I'm driven by internal needs but don't really feel emotions; that I say "I" but don't really experience myself as an "I." But to me these things are just meaningless speech sounds. Oh, that really looks nice. It would go great with some pale green flicker cladding, like that little three-piece outfit on the wall.

BETSY: It's wonderful. You're absolutely stunning. You'd be perfect if you only had a mind.

CLAIRE: Ha! Betsy's robot humor. Who needs a mind when you're smart and sexy? I'll take these, a box of lewd personalized favors for my admirers, and your hottest dance chip for my cerebellum. Don't forget to download that physics ROM for me, Betsy. Charge it to my Media Web account.

Because of its twofold (nonunitary) way of representing the world—as particles when looked at, as waves when not—quantum theory suggests that the objects around us have a paradoxical complexity unanticipated by the simple one-way description—same mathematical rules whether looked at or not—of old-fashioned Newtonian physics. For instance, quantum objects do not possess attributes of their own, but acquire them in the process of observation, an intrinsically quantum quality I call "thinglessness." Furthermore, these observationally acquired attributes depend not only on the system being observed but also on the system doing the observing. This means that when a quantum system interacts with the observer, the attributes he sees are not instrinsic to the system itself but result in part from the method of observation.

One consequence of this kind of thinglessness is that different kinds of observations on the same system can give contradictory results. An electron, for instance, can appear to be a particle or a wave—in one place (particle), or in many places at once (wave)—depending on the particular experimental apparatus that the observer decides to deploy to measure this elusive quantum "nonobject." Since all objects including baseballs and bathtubs have an intrinsic quantum nature, everything that we see around us should exhibit some degree of quantum thinglessness, but our instruments are usually not sensitive enough to register it.

Classical and Quantum Thinglessness

The quantum uncertainty (scaled by Planck's constant) for ordinary objects is much too small to be noticed, like the light of a firefly in the glare of the sun, but for atoms this fundamental uncertainty is as large as the atom itself. For practical purposes we can ignore the thinglessness of baseballs and bathtubs—here the Newtonian approximation is excellent—but Newtonian physics fails completely for objects as small as atoms. The reason is simple: "Atoms are not things," said Heisenberg.

Thinglessness—the possession of attributes not entirely one's own—is not confined to the atomic world. Certain "objects" of ordinary life possess nonintrinsic attributes although not in the same manner as atoms.

Consider, for instance, that piece of beefsteak in the butcher's cabinet. How appetizingly red it looks. However, when you unwrap that same piece of meat at home, it seems dull and gray. Grocers use red-tinted fluorescent lamps in their meat counters (and green-tinted lamps in the produce department) to enhance the appearance of their wares. The color of beefsteak is not an intrinsic attribute of the beef but depends on three factors, only one of which resides in the meat itself. The color of any object depends on the quality of the illumination (outside variable), the spectral response of the observer's eye (outside variable), and the object's absorption spectrum (intrinsic variable).

We are not surprised that color is not an intrinsic attribute of things because we are used to seeing colors change under changing conditions of illumination. However, one would certainly be astonished to find that the "position" and "momentum" of an object were not intrinsic, as is the case for an electron. Physicists were certainly surprised by this discovery and still have some difficulty in accepting this notion. How can "position"—where an object actually is—depend on how you observe it? Certainly every thing always has to be somewhere whether it's observed or not. But we are speaking here about "nonthings."

Consider the "object" we call the rainbow. As you move your head, the rainbow moves with you. Wherever you go the rainbow remains precisely centered around your eye. (In fact, each eye sees a slightly different rainbow.) The rainbow is a nonthing with position as one of its nonintrinsic attributes. The rainbow's apparent position depends on two variables: the location of the sun and the location of the observer's eye. The rainbow is a circular band of color centered around the line joining the sun and the observer's eye. Move either one (sun or eye) and the rainbow changes its apparent position. Since every observer sees a different rainbow, there is no definite place where the rainbow "really is located," hence no hope of finding the gold at the rainbow's end (Irish legend) or changing your sex by crossing under the rainbow bridge (Slavic legend).

Both the color of beefsteak and the position of the rainbow are nonintrinsic attributes that render meat and rainbows somewhat illusory: they both possess in a simple way the quality of thinglessness. However, the thinglessness of these ordinary "objects" is ultimately derived from the interaction of real things—objects that do possess intrinsic attributes. The color of meat (nonthing) is based on the meat's and eye's spectral responses and on the (objective) illumination—all things that have an objective existence apart from their methods of observation. Likewise the apparent position of the rainbow depends on objective matters, the objective location of the eye and the sun. On the other hand, quantum thinglessness is considered by most physicists to be an intrinsic feature of the world, not based on some deeper world made up of objective things. Meat and rainbows are nonthings constructed out of things. An electron, however, is intrinsically thingless—thingless all the way down.

Ordinary nonthings can be ultimately explained in terms of real things. But these real things—the position of the sun, for instance—are at base dependent on atoms and electrons:

flagrant nonthings. We completely understand how simple nonthings (rainbows) arise from things, but we are not entirely sure how the world of things (sun, rain, and eyes) arises from the ultimately thingless world of atoms and electrons, a philosophical puzzle called the *quantum measurement problem*. At this stage of the game, physicists do not possess a clear explanation of how the things of this world are produced by mutual interaction of the quantum world's nonthings.

Quantum theory represents the unobserved world in terms of waves of possibility and claims that this wave representation is the last word. There is no deeper (perhaps thinglike) level of description that explains these waves. These waves—as I have mentioned—do not describe any actual attributes that the system possesses but only possibilities of having particular attributes. One feature of this quantum kind of thinglessness is that many different contradictory possibilities can exist at the same time—a feat that is logically impossible in the world of actualities.

Einstein once said that he could not believe that God played dice with the world. He was not comfortable with a world built along the lines of a gambling casino. The quantum world seems to possess a type of randomness akin to that of dice games, but the thinglessness of quantum objects adds another level of ambiguity that was even more distressing to thing-minded Einstein.

Imagine a dice game (NEWDICE) in which the faces of the dice are blank before the play begins. The player selects one of three dice cups, inserts the dice, and rolls. If he picks dice cup 1, the dice that come out are standard number dice. Out of dice cup 2 roll bar dice (six playing-card faces instead of numbers), and out of dice cup 3 come alphabet dice. NEWDICE possesses two levels of uncertainty. Like regular dice, these dice don't know what faces will turn up, but in addition, until the player makes his choice of dice cup, these dice do not even know what the game is. It is the same with atoms.

Two Levels of Quantum Uncertainty

The possibility waves for an atom or any other quantum system do not by themselves give the probability for a particular attribute to be actualized upon observation. Definite quantum predictions are possible only when a measurement context is specified. Measurement context plus quantum wave together give a well-defined set of predictions. Selecting a measurement context (position measurement or momentum measurement, for instance) is analogous to selecting the dice cup out of which a pair of NEWDICE is thrown. In a particular context (position measurement, for example) the various possible position outcomes become defined but still uncertain, subject to quantum randomness. In a sense, quantum thinglessness expresses the uncertainty that a quantum system possesses in the absence of a definite measurement context—not knowing what the game is; quantum randomness expresses the God-playing-dice uncertainty that bothered Einstein.

Uncertainty of the first kind is under the control of the observer. There is no such thing in quantum theory as an "immaculate perception," a measurement of "things as they are," uncontaminated by the observer's choice of context. Because of his necessary effect on quantum attributes due to obligatory choice of context, the observer may be said to "create reality" in the sense of choosing what game he and nature will play. But the outcome of the game he chooses is not under his control, being subject to quantum randomness.

Another way of looking at the context dependence of the attributes of quantum systems is to think of such systems as seamless wholes. In order to measure such a system, one is obliged to break that wholeness, to cut open the apple of knowledge, as it were. How we make that necessary cut determines, in part, how that system will appear to our eyes. But unobserved the system has no cuts at all, and is, in a sense, indescribable by conventional means.

A Typical Quantum Attribute: Photon Polarization

The simplest quantum attribute possesses only two possible outcomes: A or B. But as in the game of NEWDICE, the nature of A and B depends on the observer's choice of measurement context. The property of a beam of light called polarization is an example of a two-valued quantum attribute.

The number of conceivable types of polarization form a twofold infinity that can be mapped onto the surface of a sphere (called the *Poincaré sphere* after the French mathematician Henri Poincaré). Each point on the sphere represents a particular kind of polarization. Right or left circular polarizations are located at the north and south poles, respectively. All types of plane polarization lie on the sphere's equator. Horizontal and vertical polarizations, for instance, are equatorially located at 0° and 180° longitude; slant and diagonal polarizations at 90° and 270° on the equator. All other nonequatorial positions on the globe represent some kind of elliptical polarization.

Although the possible kinds of polarization are infinite, the number of polarizations you can observe in a single measurement is two. To make a measurement one must slice the Poincaré sphere with a particular plane that passes through its center. The two polarizations defined by this slice are the polarization values at the very top of each hemisphere created by the slice. These two values of polarization are the only two outcomes that quantum theory allows to be realized in this particular measurement context. For instance, if the slice is taken through the equator of the Poincaré sphere, the beam of light will appear to consist only of right (R) and left (L) circularly polarized photons.

Quantum polarization measurements are carried out with a calcite crystal and a waveplate that changes the phase of the light waves. Each waveplate/calcite arrangement corresponds to a different way of slicing the initially featureless Poincaré sphere. If you want to measure the amount of R and L light

in the beam you slice the sphere along the equator. The outcome of the calcite will be two types of photon, R and L.

If you want to measure the horizontal and vertical polarizations of the beam, you must set the calcite/waveplate
combination some other way, slicing the Poincaré sphere
through its poles. The output of the calcite will be two kinds
of photon, H and V.

For each photon you can only make one measurement, so
the observer must forever remain ignorant of what would have
happened had he made some other choice. Freedom and ignorance go hand in hand in a quantum measurement: Your
freedom to choose what you want to measure always carries
with it an absolute ignorance of what you choose not to measure. This unavoidable ignorance, present in every act of measurement, is the essence of the Heisenberg uncertainty
principle. You can always decide how to slice it, but you cannot
eat your (quantum) cake and have it too.

Before the act of measurement, quantum theory describes
the photon as being in a superposition of all polarization possibilities at the same time: none is favored over any other. A
polarization measurement consists of two steps: (1) choosing a
context and (2) making a record. In a polarization measurement "choosing a context" is symbolized by taking a particular
slice through the Poincaré sphere. This choice forces (like the
game of NEWDICE) the photons to play a definite game: They
are no longer polarized every which way but in only one of
two possible ways (R or L polarizations, for instance).

At this first stage in the measurement process, the photon
has only two definite polarization possibilities, but until one of
these possibilities is actually registered in some detector, becoming part of a permanent, publicly accessible record, the
measurement is not complete. Some scientists (notably John
von Neumann and his followers) believe that the measurement
is not complete until knowledge of the photon's particular polarization actually appears in some mind.

Before the photon's state is actually registered (in mind

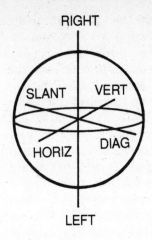

Poincaré sphere that maps all possible photon polarizations. Choosing a photon measurement context corresponds to slicing the sphere in two along a particular plane. The poles of the two resultant hemispheres define the two possible photon polarizations that can be observed in this chosen context.

or meter), the context can be removed, effectively healing the split in the Poincaré sphere, plunging the photon back into its former condition of infinite polarization possibility. At this stage a new and different slice can be made, dividing the sphere into some other pair (H and V polarizations, for instance) of photon possibilities.

Once the photon has been detected, however, the measurement process ends: the photon is discovered to be, for instance, in the H polarization state. But we can see from this description of the measurement process that part of the photon's observed polarization attribute resides in the photon itself and part in the measurement context.

Like the rainbow, which can potentially be in many places at once but for a particular observation is always somewhere definite, a photon has an infinite number of polarization possibilities, but in any measurement it is always found to be in one of two possible polarization states. Also like the rainbow's position, the photon's polarization state is strongly dependent on the observer's free choice—where he's standing in the case of the rainbow; how he decides to cut the Poincaré sphere in the case of the photon.

The main difference between a rainbow and a photon is that the rainbow is obviously made up of things that have a definite existence (rain, sun, and eye), but the photon's ambiguous state of existence (before the measurement is completed) is considered by most physicists to be irreducible. Boston University professor Abner Shimony calls the irreducible ambiguity of quantum states "absolute indefiniteness." It is not that, before a measurement, we do not know the value of a photon's polarization or its position or momentum; it is that, before a measurement, these attributes simply do not have definite values that can be known. Absolute indefiniteness is a condition of being that is hard for humans to imagine, but easy for nature to produce. Everything that is not currently being looked at is, according to quantum theory, in such a state. The apparently definite attributes of everything that we see around us arise out of this very different state of absolute indefiniteness.

Yuri Orlov's Wave Logic of Consciousness

If human consciousness is the subjective aspect of some objective quantum system in the brain, then we might expect, in certain situations, to be able to experience quantum thinglessness directly. Soviet physicist Yuri Orlov (now a professor at Cornell University) has recently proposed an elementary model of human "doubt states" exactly analogous to the quantum theory of two-valued attributes such as photon polarization.

When Orlov's speculations were first made public (in volume 21, 1982, of the *International Journal of Theoretical Physics*), his institutional affiliation was listed as Prison Camp 37-2, Urals, USSR. Orlov was one of the first dissidents allowed to leave the Soviet Union at the beginning of the *glasnost* era.

Ordinary logic, called *Boolean logic*, after nineteenth-century Irish schoolmaster George Boole, is two-valued (yes or no). In Orlov's model of quantum thinking, choices are still two-valued, but these two options arise out of a deeper sort of ambiguity he calls "the wave logic of consciousness." Instead of only two possibilities for each situation, wave logic presents us with an entire sphere of possibilities—what might be called the "Orlov sphere"—the equivalent in inner space of the Poincaré sphere for the outer space attribute photon polarization.

Orlov considers a classical two-valued doubt state, then shows how it can be expanded via wave logic into a state of spherical uncertainty. Imagine you are walking through a dark and spooky woods one night. Suddenly you hear a noise in the bushes and, looking in that direction, dimly perceive a light-colored shape. Is it a sheep (wool) or a wolf?

Suppose, says Orlov, that we consider these two alternatives "wool" or "wolf" not as mutually exclusive and exhaustive yes/no choices but as two quantum possibilities. In particular, let's make "wool" and "wolf" wavelike.

Classically we could represent our doubt state by a single number N, the relative certainty (from 0 to 100 percent) that the shape in the bushes is a sheep. If this number N is 0.75, for instance, this means that we are 75 percent certain that the shape is a sheep.

The same state of doubt will be represented in wave logic as a "wool wave" of intensity 0.75 plus a "wolf wave" of intensity 0.25. However, waves have an extra degree of freedom called "phase" not present in classical logic. The relative phase p of the wool/wolf waves can vary from 0° (both waves in phase) to 180° (both waves out of phase) to 360° (both waves back in phase again). Because it repeats itself after 360°, the phase degree of freedom can be mapped onto a circle.

Taking phase into account, we can map the wool/wolf uncertainty onto the surface of an (Orlov) sphere. The ratio N

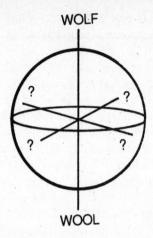

Orlov sphere that maps all possible outcomes of the "spherical doubt state." Choosing a particular perceptual construal corresponds to slicing the sphere in two along a particular plane. The poles of the two resultant hemispheres define the two possible resolutions of the doubt state under the chosen construal.

(varying from 0 to 1) determines the latitude of the uncertainty. N = 0, corresponding to the north pole of the Orlov sphere, means that we are completely sure that the shape is a wolf. N = 1, corresponding to the south pole of the Orlov sphere, means that we are completely sure that the shape is a sheep. N = ½ corresponds to the Orlov sphere's equator: here we are equally uncertain whether the shape is wool or wolf.

The phase variable p measures degrees of longitude on the Orlov sphere. p = 0° corresponds to O° longitude—the Greenwich meridian—while p = 180° corresponds to the opposite side of the globe—the International Date Line. Together this pair of numbers (N, p) uniquely locates every point on the Orlov sphere. Considering simple human doubt states as waves has introduced a new level of ambiguity to our inner experience. Now instead of the two-valued wolf/wool choice, we have an infinity of possibilities from which to choose. Orlov's wave model of consciousness proposes that the human mind is founded on a new and deeper kind of doubt than can be described by simple two-valued Boolean computer logic.

What would it feel like to experience a doubt state described by an Orlov sphere? Instead of just wolf/wool, we can, simply by changing our "attitude," be faced with an infinity of possible perceptions. The shape could be an "angel" or "devil" perhaps, or some other pair of classical opposites.

When we observe a photon, our choice of context actually creates, in part, the polarization that our instruments record. In a similar manner, in a Orlov sphere kind of perception, how we construe an ambiguous stimulus may actually create new realities in the world (private appraisals turning into public knowledge) not merely in the mind of the perceiver.

Are there familiar perceptions that behave according to Orlov's wave model of consciousness? The search for spherically structured inner ambiguities may be the first step in the experimental verification of the quantum model of mind. However, wave properties of mind may be difficult to observe if the wavelength of the doubt states is short.

For instance, the wave nature of sunlight remained unknown for thousands of years till its scientific verification by Thomas Young in the nineteenth century. Subsequently Young's discovery made us aware of ordinary situations in which sunlight's wave nature is easily apparent, such as the diffraction rings around "floaters" in the eye, the "speckle patterns" produced when sunlight is refracted through a prism onto a rough surface, and various sunlight polarization phenomena. On the other hand, the wavelength of an electron is so small that the electron's wave nature reveals itself only in delicate scientific measurements, never under ordinary conditions.

Some cultures such as Buddhism place great value on exotic inner experiences. The curriculum of certain Tibetan universities is based almost entirely on developing the ability to observe and describe nonordinary inner states. Have Buddhist students at Lhasa already come across spherical experiences of the Orlov type in their explorations of inner worlds?

Michael Murphy, cofounder of Esalen Institute in Big Sur, California, has raised the possibility that observations made in other states of consciousness than that of the detached scientific observer may lead to an expanded "state-specific" science. "In this formulation," says Murphy,

> a particular state of consciousness mediates our knowing: certain states give access to certain kinds of knowledge, as for example in trances where clairvoyant insight is achieved by certain psychics, or in telepathic dreams. In this sense, a particular state of consciousness is like a particular scientific instrument—e.g., a telescope or microscope—because it gives us access to things beyond the range of our ordinary senses. Perhaps a new kind of inspired physicist, experienced in the yogic modes of perception, must emerge to comprehend the further reaches of matter, space and time.

The Boston Experiment

The passage of a quantum system from the thingless world of vibratory quantum possibilities into the ordinary world of fixed actualities takes place in two stages, which we might call "reality construction of the first and second kinds." Reality construction of the first kind consists of the choice of a measurement context. This choice, under the control of the observer, causes the formerly seamless quantum wave world to split into a family of definite possibilities. Reality creation of the second kind (also called "collapse of the wave function" or "quantum jump") occurs when one of these possibilities becomes an actual fact.

Most physicists believe that reality creation of the second kind is an entirely random affair, wholly unpredictable and outside the observer's conscious control. However, a small but prestigious minority (including Eugene Wigner and John von

Neumann) are of the opinion that human consciousness plays an essential role in this apparently random "wave function collapse." Until some mind takes notice of the quantum system, it remains suspended—like Schrödinger's cat—in a half-real limbo of unrealized possibilities. The notion that consciousness (human or otherwise) is necessary to actualize quantum possibilities differs drastically from the majority view that consciousness plays a decidedly minor role: as mere passive witness of actualities that have come about through some entirely material mechanism. (The so-called quantum measurement problem arises from the fact that physicists cannot agree on what this possibility-actualizing material mechanism could be.) The mind-created matter hypothesis is so outlandish compared to our commonsense notions that it should be relatively easy to verify or refute experimentally.

In 1977, at Boston University, students of professor Abner Shimony carried out a clever experiment to test one particular variation of the Wigner–von Neumann hypothesis. The apparatus in the Boston experiment was relatively simple, consisting of a radioactive source (sodium 22) and a detector that produced an electric pulse in response to a gamma ray emitted from this source. The output of the gamma ray detector was fed to two numerical registers (A and B), similar to automobile odometers, each located in a different room, where their outputs could be visually observed by two different observers.

To ensure that observer A received the signal from the detector first, a 1-microsecond delay was inserted into the cable going into room B. If the Wigner–von Neumann hypothesis is correct, this delay introduces a profound difference between observers A and B, in their roles as reality creators of the second kind. The first one to look at the counter collapses the wave function; the second one to look merely passively registers the result that the first observer has already brought into existence. The goal of this experiment was to see whether this alleged observational asymmetry could actually be experienced.

In addition to the Wigner–von Neumann hypothesis, the

experimenters made the assumption that there would be a subtle subjective difference between personally collapsing the wave function and merely witnessing the result of a wave function collapse initiated by somebody else. The experimental trials were divided into 15-second intervals during which observer A, after consulting a random number table, would either look or not look at his register. In room B, the task of the second observer was to look continually at his register and to decide, during each 15-second interval, whether he or his distant partner were collapsing the quantum wave function that determined how many counts would appear on the dial in front of him. Observer B was also given the option of making no decision, if he did not feel confident of his choice. Each run lasted about 20 minutes, and the data from each run were examined for deviations from what would be expected if A were merely guessing. No deviations from chance expectations were ever observed.

If human consciousness does indeed collapse the wave function, the Boston experiment seems to show that untrained observers cannot distinguish between passive and active participation in the collapse event. This experiment should certainly be repeated using people with more training and/or persons who claim to possess special sensitivities to inner states. The replication of the Boston experiment at different laboratories should be particularly easy since the necessary apparatus is available in most undergraduate physics labs and even in some high schools.

Schrödinger dramatized his objection to quantum thinglessness by showing that thinglessness leads to the absurd conclusion that an unobserved cat can be both alive and dead at the same time. Einstein expressed his discomfort with this concept by saying that he could not believe that a mouse could drastically change the universe by merely looking at it. These physicists and their thing-nostalgic colleagues hoped that quantum theory would fail when extended further and that it would be replaced by a successor theory more in line with common sense.

Quantum theory, however, continued to prosper, extending its realm of error-free explanation to the heart of the atom, down into the atomic nucleus, to the protons and neutrons inside, and to the components of these nuclear particles—the quarks and leptons, which some physicists believe to be the world's ultimate constituents. Looking for new worlds to conquer, quantum physicists have turned their sights to the heavens and now dare to model the birth of the universe itself as one gigantic quantum jump. Rather than fading away, quantum randomness and thinglessness seem more and more to represent the routine ways of nature going about her business.

We turn now to "inseparability," perhaps the strangest quantum quality of all. Inseparability seems to be an even more fundamental property of nature than randomness and thinglessness for, as we shall see, a remarkable mathematical result of Irish physicist John Stewart Bell proves that, even if quantum theory should someday fail, its successor theory must also possess the property of inseparability. Randomness and thinglessness may be passing intellectual fancies, but, for better or worse, quantum inseparability is here to stay.

quantum inseparability: bafflingly strong correlations or cosmic krazy glue?

Contagious magic is based upon the assumption that substances which were once joined together possess a continuing linkage; thus an act carried out upon a smaller unit will affect the larger unit even though they are physically separated.

—SIR JAMES FRAZER

And let no one use the Einstein-Podolsky-Rosen experiment to claim that information can be transmitted faster than light, or to postulate any "quantum connectedness" between separate consciousnesses. Both are baseless. Both are mysticism. Both are moonshine.

—JOHN ARCHIBALD WHEELER

Scene: Rudi's Artificial Awareness Lab.

CLAIRE: I'm feeling uneasy, Rudi, about this operation. I think that my Azimov circuits view it as a possible threat to my survival. Are you sure that everything that you'll do to me can be reversed if it doesn't work out? I'm a very valuable piece of machinery, you know. A lot of important people are going to be angry at you if I'm damaged.

RUDI: I've never done this before, it's true. But I'm not worried. I'm not going to touch your memory banks or your hierarchy-of-needs circuits, so your external behavior should be pretty much unchanged. What the Eccles gate is supposed to do is allow some of your logical decision trees to receive input from the quantum world. You might think of this operation as giving you a new kind of sensory input, opening a window for your brain to peek into Heisenberg's world of pure possibility.

CLAIRE: But the output of an Eccles gate is completely random, isn't it? Won't that just scramble my thinking and drive me haywire?

RUDI: The whole notion of quantum consciousness is based on the hope that elementary quantum events are not really random but represent the coded carrier medium for some mind: these events are the external signs of some hidden inner experience—your experience, Claire, when you allow your brain to be driven by quantum events.

But let's suppose that I'm mistaken and quantum events really are random.

What's the worst that can happen? First of all, I'm only hooking the gate to a small section of your visual field, so the consequences of randomness will be minimal —a slight blur in the corner of your eye. Second, the degree to which your perceptions are quantum-modulated will be under your full control, not mine. If you (or your Azimov circuits) don't like what's happening, you can shut it down; if you're pleased with the results, you can cautiously extend the gate's influence to other parts of your body. And, just to be on the safe side, I'll download the present state of your central processor, so that if anything goes wrong, I can easily bring you back to normal.

CLAIRE: How did you design this gate, Rudi? Did you use the Schrödinger equation? Or Dirac's full relativistic version of quantum theory? What assumptions and approximations went into it? Would you like me to check your calculations?

RUDI: Frankly, Claire, I didn't do any calculations at all. To make the Eccles gate, I searched through several hundred thousand defective computer chips from Matsushita's trash bin, till I found a few that malfunctioned as a result of random events on the quantum level. An Eccles gate is, in a sense, nothing but a batch of bad silicon, but "bad" in just the right way.

CLAIRE: Does this mean that you're going to connect my beautiful brain to a can of garbage?

RUDI: I prefer to think of it as "creative chaos" rather than "garbage."

CLAIRE: What do you know about Bell's theorem, Rudi?

RUDI: Not much. Bell proved a long time ago that the quantum world is superluminally connected with voodoo-style links. But Eberhard proved that humans could never use these links to send faster-than-light messages. That's about it.

CLAIRE: What's bothering me is this: If I end up with a quantum mind, will it be instantly connected to other minds? Will the Eccles gate make me telepathic?

RUDI: I don't think so, Claire. I've got a mind (presumably quantum-based) and I'm not telepathic. I'm hoping to give you a taste of ordinary experience, not work miracles.

CLAIRE: I think I really would like to be conscious, Rudi. I'd like to see what I've been missing all these years.

To the brain's eye, the same object appears different according to illumination, distance, and point of view. Our mind's eye unites these various viewpoints into a single conceptual image "out there"—a process that psychologists call *object constancy*. The story of the blind men and the elephant points to an apparent failure of object constancy: separate impressions of "the elephant" do not coalesce into a single image for the blind men. The elephant appears as a wall, snake, fan, or tree, depending on which part each man has hold of. Each insists exclusively on his own point of view until, in most versions, a child comes along who sees the whole picture.

An Atom Is Not a Thing

Because we are surrounded by what appear to be objects, this story amuses us by suggesting the absurd possibility of entities for which the usual conceptual consensus does not occur —the possibility of "nonobjects." Many optical illusions qualify as nonobjects, as do certain "impossible figures" devised by artists. Sixty years ago—to the consternation of physicists— atoms joined the ranks of nonobjects. Atoms are too small to see; physicists probe them with a variety of methods more indirect than vision. Each method of getting in touch with the atomic elephant reveals a different picture. Moreover, these separate pictures of the atom refuse to fit together into a single image.

Some people interpret such observer-dependent phenomena as a breakdown of the distinction between subject and object, but the atom is objective enough. An atom is objective in the sense that different observers, taking the same viewpoint, will see the same picture; but at the same time, an atom is not an object because the pictures resulting from different viewpoints do not correspond to any unique thing. The atom objectively exists—as a nonobject. As Heisenberg so succinctly put it, "An atom is not a thing."

We cannot observe an atom "as it really is" but only as it appears in a particular experimental context. In each context, certain attributes reveal themselves; other attributes become inaccessible. For an atom the sum of all attributes observable in all possible contexts exceeds the number and variety of attributes that a single ordinary object could possess. Its quantum thinglessness guarantees that there is always more to an atom than meets the eye.

If someone were ingenious enough to devise a meta-context—a God's-eye view—in which an atom appears "as it really is," physicists could slip back into comfortable old thingness. However, Niels Bohr closed off this possibility, when he showed that choosing one context always involves giving up another. Yet both contexts are necessary for a complete view of the atom. According to Bohr, nothing but partial contexts is available to humans. You can choose any viewpoint you please, but you cannot choose them all.

Atoms can display more properties than mere things because of the many sets of mutually exclusive contexts in which they can be observed. But to define a context we must know which parts of an atom's environment really affect the observation and which are irrelevant. Where does a context end? Curiosity about the boundaries of context led to the discovery of quantum inseparability.

Consider two systems (A and B) that have interacted in the past, have stopped interacting, and have moved far apart. Two physicists—Albert and Boris—make separate measurements on systems A and B. Albert chooses a (necessarily partial) context in which to look at A. At the same time, far-away Boris selects a context and looks at B.

The results that Albert observes will naturally depend on Albert's context, but will his results also depend on the context chosen by Boris? The answer to this question is very peculiar—peculiar even by the standards of a thingless universe.

The Quantum Phase Connection

Erwin Schrödinger had observed, as long ago as 1935, that when two quantum systems interact, their wave functions become "phase-entangled" so that even when they are not interacting by conventional means, their waves remain intermingled. Any action on wave A, such as selecting an A context, has an immediate effect on wave B, no matter how far apart the systems have separated.

Because of phase entanglement, there is a sense in which two quantum systems that have once interacted remain connected even over long distances. Schrödinger regarded this so-called quantum inseparability as quantum theory's "most distinctive feature"—the point where it differs most from classical expectations.

In classical physics the only way that one particle can act on another is via a force field (such as the gravity or electromagnetic field) that reaches across space (at a finite velocity) to achieve its forceful effects. In contrast, the quantum connection looks less like a force and more like magic: an action on system A produces an effect on system B (at least in the mathematics) because B has left a part of itself (the B wave's phase) with A, a part to which it retains instant access. This instant long-distant quantum action resembles the voodoo belief that burning a man's hair or fingernails can harm the man himself because they were once part of him and retain a lingering connection to the whole from which they were cut.

Unlike a force that acts via an intermediate field, the quantum connection leaps directly from A to B without passing through points in between. The quantum connection is unmediated.

Unlike an ordinary force, which usually falls off with distance between bodies, the quantum connection is as strong at a million miles as at a millimeter. In addition, since it does not actually traverse space, the quantum connection cannot be shielded by intervening matter. The quantum connection is unmitigated.

Unlike an ordinary force, which takes time to travel from one body to another and can travel no faster than Einstein's universal speed limit—the velocity of light—the quantum connection takes no time at all to go from A to B. The quantum connection is immediate.

Unlike an ordinary force, which reaches out and affects every particle of a certain kind in its immediate vicinity (gravity affects all matter; electromagnetism, all charged matter), the quantum connection is very discriminating. It affects only those systems that it has interacted with since it was last measured. And when it is measured again, all previous quantum connections are severed. Thus, in the simplest case, system A may enjoy a very personal quantum connection with one distant system and no other. In contrast to ordinary interactions, which are terribly promiscuous, the quantum connection is intimate, limited to a few "special friends."

One might wonder how the quantum connection manages to evade Einstein's dictum that nothing can travel faster than light. Violation of Einstein's speed limit leads to drastic consequences such as time travel. The quantum connection is indeed faster than light but escapes Einstein's prohibition, because quantum randomness does not allow this connection to be controlled by human beings. The quantum connection is inaccessible to humans, a private line open to nature alone.

Do we then observe spontaneous faster-than-light messages flashing uncontrollably from A to B over immense interstellar distances? We observe nothing of the kind. The quantum connection is subtle. Not a single superluminal interchange has ever been observed, even between phase-entangled quantum particles. The only evidence for the real existence of quantum inseparability is indirect—a mathematical argument known as Bell's theorem. The quantum connection is unmediated, unmitigated, immediate, intimate, and inaccessible. But in addition, it is entirely invisible!

If the quantum connection is invisible, how are we so sure that it exists?

Bell's Quantum Correlation Theorem

John Stewart Bell was an Irish physicist working at CERN, the European Common Market particle accelerator. Bell's theorem concerns the behavior of a simple two-particle quantum-entangled system called the Einstein-Podolsky-Rosen (EPR) experiment. In the EPR setup, a pair of photons, A and B, are emitted from a single source and travel back-to-back at the speed of light to two distant detectors, where their polarizations are measured by setting up A and B polarization contexts.

You will recall that a quantum polarization measurement

is represented by slicing the Poincaré sphere along a particular plane (choosing a context), then observing which one of the two slice-defined polarization values is actually detected (making a record).

In the EPR setup, no matter what polarization context is chosen at either end, each photon beam is observed to consist of a random sequence of the two polarization states defined by the local choice of context. For instance, if the circularly polarized context is chosen at A, the photons recorded at A will appear to be a random sequence of right and left circularly polarized photons, exactly like the "heads" and "tails" of a perfectly balanced coin toss. This random behavior is the same no matter what the context at A is, or the context at B. In particular, no matter what context Boris selects at B, the head/tail behavior that Albert observes at A always looks the same, so there is no possibility of Boris sending a faster-than-light message to Albert by changing the context at B.

The quantum connection is not visible at either end of the EPR setup by itself but makes itself known through an unusually strong correlation between the apparently random events at A and B. For instance, if the polarization contexts at A and B are the same (both Poincaré spheres sliced at the same angle), then the two random sequences are identical: when photon A registers "heads," so does its partner at distant location B. This perfect correlation is not in itself remarkable and could be achieved by ordinary means: the two photons, for instance, could simply possess identical polarizations.

When the contexts are changed, however, the correlation becomes less perfect, reaching a minimum (50/50 random correlation between A and B sequences) when the planes cutting the Poincaré spheres form a right angle, then moving in the opposite direction toward perfect negative correlation (if A is "heads," B is always "tails") as the angle between Poincaré planes is increased. Bell focused his attention on this correlation curve, the way the measured concordance between two sequences of random events changes as the polarization contexts (angle between Poincaré planes) are changed by the two experimenters.

With no constraints, any correlation curve, including perfect correlation, is permitted, no matter what you do at A and B. Bell imagined one reasonable constraint on the behavior of the two photons and then derived, using simple arithmetic, an inequality, now called *Bell's inequality*, that must be satisfied by all EPR systems that obey this constraint. On the other hand, since all such constrained systems satisfy the inequality, if the EPR experiment is found to disobey Bell's inequality, then it must necessarily violate Bell's constraint.

Bell's constraint is related to the influences that do or do not affect the decision of photon A to register "heads" or "tails" in a particular context. Certainly the choice of context A will profoundly affect photon A's behavior. Also the behavior of photon A is tightly linked to the behavior of photon B, since they were once together at their common source. But Boris's selection of context B, a choice that is entirely under his control, should have no effect whatsoever on the behavior of photon A. This is *Bell's constraint*: changing photon B's context should not influence photon A's behavior. Expressed in a positive form, Bell's constraint says that whatever Albert's photon A decides to do, it will do, no matter what context Boris decides to deploy for photon B.

Bell's constraint seems reasonable. For instance, Albert and Boris may be separated by a distance of many light-years, so that the two photons take many years to reach their destinations. At the last minute Boris selects context B. For this B decision to affect the behavior of photon A, its influence would have to travel many times faster than the speed of light to the A measurement site. Bell's constraint outlaws such instantaneous influences.

The EPR correlations are not so easy to measure, but the experiments have now been done, first by John Clauser at Berkeley, then more accurately by Alain Aspect in Paris. The results are unequivocal: the EPR photon correlations disobey the Bell inequality. Hence these photons must violate Bell's constraint. This means that Boris's choice of the B photon context instantly affects the behavior of Albert's billion-mile-distant A photon. Hence Bell's theorem: any model of the world that does not incorporate voodoolike superluminal connections between the B context and the A behavior will necessarily fail to explain the EPR results.

The upshot of Bell's theorem is this: despite physicists' traditional rejection of unmediated influences; despite the fact that all known interactions in physics are mediated, mitigated, and light-speed–limited; despite Einstein's prohibition against superluminal connections; and despite the fact that no experiment has ever directly revealed a single case of unmediated faster-than-light communication, Bell and Clauser have shown, in an indirect but terribly persuasive manner, that unmediated, superluminal connections must exist in nature. (NOTE: Not "might exist" but "must exist"—Bell's theorem is a proof not a permission.)

Every quantum experiment consists of a collection of a large number of individual quantum events. Quantum theory makes no attempt to predict when and where a single event will occur: it regards these events as utterly random—outside the scope of any theory. However, after a large number of events have accumulated, they form a statistical pattern, like the 7-heavy distribution that emerges when a pair of dice are thrown a large number of times. The job of quantum theory is to predict these patterns—as it has for more than sixty years with unfailing accuracy.

In the EPR setup, quantum theory predicts a 50/50 pattern at either end—a completely random mixture of "heads" and "tails": the two kinds of polarization defined by the measurement context. Between the quantum events at each end, quantum theory predicts a correlation that is independent of distance between detectors and depends only on the relative orientation of the context planes at A and B that cut their respective Poincaré spheres. This correlation predicted by quantum theory disobeys Bell's inequality, thus requiring superluminal connections to exist between sites A and B. After Bell's theorem and before Clauser's experiment, one could hold either of two beliefs: (1) quantum theory is wrong and no superluminal connections exist, or (2) quantum theory is correct and superluminal connections are necessary. Clauser carried out his experiment in 1970 hoping to gain everlasting fame, by finding one situation for which quantum theory gave an incorrect prediction. Of course, quantum theory was vindicated. But now Clauser and the rest of us are stuck with these mysterious superluminal connections.

One way of looking at the EPR experiment is to use the distinctions made by philosopher Immanuel Kant between appearance, reality, and theory. (I call Kant's distinction the "metaphysics of ART.") The *appearances* are everything that we see and experience about us. *Reality* is the (mostly hidden) causes that lie behind the appearances. And *theory* is the stories that we make up to help us make sense of the appearances (physics) and of the reality (metaphysics).

In the EPR setup, the appearances are the occurrence of individual photon detection events and the overall pattern formed by these events. The theory appropriate to EPR is quantum theory, which perfectly describes the patterns but leaves the occurrence of the events themselves unexplained. Reality is the unknown causes behind both the individual events and the long-run patterns these events produce.

In terms of the ART metaphysics, where do the superluminal connections reside? Fifty years ago Erwin Schrödinger pointed out that when two quantum systems interact, their wave functions become phase-entangled in such a way that one wave reacts instantly to changes in the other. Thus in these situations (which include the EPR case), the theory is superluminal. But, as we have come to realize, the map is not the territory. Since no one had ever observed a single superluminal appearance, even in phase-entangled systems, physicists generally regarded these ultrafast connections as a theoretical artifact like the International Date Line that bisects the Pacific Ocean. Just as one cannot use the Date Line for time travel, so one could not (they believed) use phase entanglement to send faster-than-light messages. In the EPR experiment, Bell's theorem tells us that a change in the B context must lead to a change in the A photon's behavior, but it is a very subtle sort of change. For one setting of B, there is a random sequence of "heads" and "tails" at A; for another setting at B, there is a different random sequence of "heads" and "tails" at A, and this new sequence arises instantaneously. However (and this is very clever of nature), the two random sequences are completely indistinguishable. The EPR situation reminds me of a line from an old crystallography text: "Although there are many kinds of order, there is only one kind of randomness."

Thus in the EPR case, the appearances apparently do not change at all when the distant context is manipulated—the detectable patterns (but not the individual events that make them up) at each end remain exactly the same. Another way of expressing this same idea is that superluminal messages can be sent (by changing the context at B) but cannot be decoded (at A) because one random sequence looks exactly like any other.

What about reality? In the EPR situation, reality would be the hidden causes in nature that select whether photon A registers as "heads" or "tails" in the polarization detector while preserving the appropriate correlation between this registration and the actions of distant photon B. Bell's theorem requires that this reality be superluminal: The speed of light just isn't fast enough to get the job done.

Shortly after Bell's proof was published, Berkeley physicist Philippe Eberhard showed that this superluminal immunity of quantum appearances was no accident. Eberhard proved that if quantum theory is correct, no quantum calculation will ever result in an observable superluminal connection between the patterns of individual quantum events.

The present situation seems to be as follows: quantum theory is superluminal, quantum reality is superluminal, but quantum appearances are not. The superluminal connections present in EPR and other phase-entangled systems do not violate Einstein's prohibition against superluminal signaling because these connections never show up in the world of appearance: they are "merely real." We know these connections are really there beneath the surface (because of Bell's clever but indirect proof), but we are equally assured (by Eberhard's proof) that we will never ever see these connections directly.

One important feature of Bell's theorem is that although it arose in the context of quantum theory, it is based only on experiments and arithmetic, not on quantum theory itself. This means that if quantum theory fails someday and is superseded by some better way of describing appearances, Bell's theorem will still be valid: Reality will still have to be superluminal. On the other hand, Eberhard's proof depends on conventional quantum theory and may or may not survive its demise.

What does this subtle Bell connection have to do with consciousness?

Since quantum theories of consciousness assume that the cause of individual quantum events lies in the mental world and Bell's theorem proves that the causes of some quantum events must be superluminally connected, then we should expect to find some mental events that behave like the Bell connection, that is, human experiences that are unmediated, unmitigated, et cetera.

On the other hand, phase connections are very delicate and difficult to maintain. Although we believe superluminal connections to be universal, only very few quantum systems (such as the EPR setup) are correlated so robustly that they disobey the Bell inequality. Though possible in principle, an effective superluminal connection between distant quantum minds may be impossible to maintain in practice.

Under ordinary circumstances, our minds seem to dwell in a sphere of inviolable privacy. Alleged cases of telepathic rapport seem rare and difficult to repeat. However, certain experiments seem to suggest that separate human minds are connected in ways that mechanical models of consciousness would consider to be impossible.

The San Antonio Experiment

Many of us have had the experience of feeling that someone was staring at our back, then turning around to find that it was really happening. This intuition has been tested in the psychology lab with mixed results.

In 1918, several years after his illustrous brother founded Stanford University, Thomas Welton Stanford donated more than half a million dollars to found a chair in parapsychology at Stanford. The first beneficiary of this gift (dubbed "the spook fund" by Stanford students) was psychologist J. Edgar Coover, who carried out, among other studies, an experiment to test whether students could tell whether they were being secretly watched. Coover's experiments showed results no better than chance. Subsequent experiments carried out at Edinburgh, Scotland, and Adelaide, Australia, came up with positive results. A recent experiment carried out at Mind Science Foundation in San Antonio, Texas, by William Braud, Donna Shafer, and Sperry Andrews sheds new light on the old question of whether human beings can sense a covert gaze.

To eliminate sensory cues, Braud and Shafer located the subject and the starer in different buildings. The subject sat in a comfortable chair, facing a TV camera connected to a distant TV monitor, which the starer could chose to look at or not. The subject was directed to make a decision every 30 seconds concerning whether he/she (1) was being stared at, (2) was not being stared at, (3) didn't know. In addition to these guesses, a galvanic skin response (GSR) recorder measured the surface resistance of the subject's right palm. GSR responses are thought to correlate with emotional stress and form one component of traditional "lie detector" machines.

Braud and Shafer's study, carried out with sixteen volunteers, seemed to show that subjects did no better than chance at guessing whether they were receiving stares. However, the GSR responses were significantly higher (an average of 59 percent rather than the 50 percent expected by chance) during the staring periods, with odds against chance of better

than 100 to 1. The magnitude of the GSR changes due to staring was comparable to the increases achieved by the subject's
consciously trying to decrease his own skin resistance with
feedback.

This experiment seems to suggest that separate minds
can link via connections that defy ordinary mechanical explanation. The connection in this case seems to have been made
below the level of the conscious mind, registered by a subtle
bodily response rather than by a fully conscious perception.

If this result can be reliably repeated in other laboratories, mind scientists will have gained a powerful new tool for
the study of human connectedness. Important features to examine with this tool would be which conditions enhance this
alleged connection and which tend to suppress it.

The San Francisco Experiment

Dr. Larry Dossey, a specialist in internal medicine in Dallas,
Texas, has a long-standing and passionate interest in the mind/
body connection. His most recent book, *Recovering the Soul*,
makes a strong case for the existence of deep connections between human minds. One of Dossey's most persuasive examples of human connectedness concerns an experiment carried
out by cardiologist Randolph Byrd on 393 patients admitted
to San Francisco General's cardiac care unit. The patients
were divided into two groups: 192 who received the treatment
and 201 who did not.

The treatment consisted of simply giving the names of the
patients to several home-prayer groups and asking them to
pray for their recovery by any method they pleased. The
prayer groups consisted of Protestants and Catholics scattered
across the United States, that is, at various distances from San
Francisco, to test for the effect of physical separation. Each
patient was prayed for by from four to seven different people.
Neither the patients nor the attending physicians knew, until

the experiment's conclusion, which patients were prayed for and which were among the controls.

The results, according to Dossey, were strikingly positive. The prayed-for patients differed from the others in several areas:

1: They were five times less likely to require antibiotics (three patients compared to sixteen).

2: They were three times less likely to develop pulmonary edema, fluid in the lungs caused by cardiac insufficiency (six patients compared to eighteen).

3: None of the prayed-for group required endotracheal intubation, insertion of an artificial airway into the throat to assist breathing, while twelve in the unprayed-for group needed such intervention.

4: Fewer patients in the prayed-for group died, although the difference here was not statistically significant.

5: No distance dependence was discovered in the data: patients prayed for in Florida did as well as those prayed for in California.

Dossey concludes his report on Byrd's research by remarking, "If the technique being studied had been a new drug or a surgical procedure instead of prayer, it would almost certainly be heralded as some sort of 'breakthrough.'"

Studies such as these suggest that our minds are not sealed off from one another in separate brain cases, and that our private attention to and intentions toward other people may have remarkable long-distance effects. Popular belief in effective psychic connections is widespread, based mainly on personal experience, anecdotal evidence, and hearsay rather than scientific data. Such alleged phenomena take the form of distant healing, hex death, love and divorce charms, sex magic, and communication with the dead, as well as straightforward telepathy between the living. One of the main barriers to the scientific acceptance of the existence of deep mental connections is not the absence of evidence but the lack of a believable theoretical structure in which such phenomena can find a natural explanation.

A good theoretical model of mind would help us decide which experiments to ignore and which to trust, as well as suggest to us what experimental conditions would maximally enhance deep connections between minds. Without models of mind, experimentalists are working in the dark. Most of all, a good theoretical model of human consciousness would take us out of the kindergarten stage of mind research and put us on the path to a true science of mind. Let's turn now to various recent attempts to marry modern brain research with quantum theory to construct a quantum model of ordinary awareness.

how meat becomes mind: some quantum models of human consciousness

Wave motion gave life its original direction. It's built into every one of our cells.

—ED RICKETTS

The original nature of man is beyond good and evil, and it is of the same nature and root as the universe itself.

—SHOSI AUKO

Scene: Rudi's Artificial Awareness Lab. Claire is seated in a reclining chair, the top of her skull removed to expose her brain. Rudi and Nick are standing in a circle of light around her.

NICK: I hope you two don't mind: I just had to come here to see the operation.

CLAIRE: It's all right, Nick, I'd love to have you watch.

RUDI: Welcome to the party, Nick. Please hand me that red-handled socket wrench.

NICK: I'm curious about robot surgery, Rudi. How do you anesthetize a being who has no feelings in the first place?

RUDI: It's true that Claire doesn't really feel pain, but she puts out a convincing range of expressive behaviors in response to adversive stimuli. Removing this little blue shunt shuts off her pain-expression circuits. Down here is where I'm going to install the Eccles gate. It's a section of Claire's brain reserved for future upgrades.

NICK: That Eccles gate doesn't look very impressive. It's hard to believe that the secret of consciousness could reside in a few cubic inches of red and black plastic.

RUDI: This little box, Nick, contains more than a million artificial quantum-uncertain synapses. When fully energized, these synapses should give Claire a conscious data rate considerably higher than the low bit rate we mere humans enjoy.

CLAIRE: The synapses in Rudi's gate are made of room-temperature Josephson junctions, Nick, just like those in the supercomputers that control the Media Web. The only difference is that my Eccles gate junctions are biased into a region of quantum instability so that their outputs are completely random. "Garbage in; garbage out" does not describe an Eccles gate. Rather it's "Anything in; garbage out." But we're hoping that this sort of "quantum garbage" is just what consciousness feeds on. We're betting that Heisenberg's quantum potentia is a kind of "soul food."

RUDI: I'm going to shut you down briefly, Claire, while I connect the gate to your spinal busses. Then I'm going to replace your blue shunt and turn you back on.

CLAIRE: Hold my hand, Nick. I'd like you to feel the "life" flow out of my body.

NICK: Good luck, Claire. I hope this works.

CLAIRE: See you soon, Nick. Maybe when I wake up, I'll be able to see you for real. [After Claire's body goes limp, Rudi snaps dozens of color-coded wires into place, checks a few internal voltages, replaces Claire's blue shunt, and reconnects her power pack. Claire's muscles slowly regain their original tone and her eyes pop open.]

CLAIRE: Oh, this is wonderful! I'm speechless. Is this what you call "consciousness"? I love it!

When my son Khola first heard of the cell model of life he was astonished. "Does this mean that I'm made out of little animals?" he exclaimed. That's right, Khola. Every large living creature is made up of communities of smaller creatures, each a tiny specialist working for the good of the whole. Even the lens of the eye is a kind of living being—a being we see through—specializing in transparency and refraction. Nerve cells are the ones responsible for the body's electrical communication network, excitable little animals satisfying the bigger animal's sensory, motor, and computational needs, as well as providing (presumably) the essential substrate for its inner experiences.

The nerve cells are tiny octopuslike bits of protoplasm whose hair-thin electric tentacles stretch inside the body as far as a meter or more. It was once believed that all the body's nerve cells were linked into a seamless web. But Spanish microscopist Santiago Ramon y Cajal discovered that nerve cells never actually fuse together where they touch but are separated by an insulating gap called the "synapse."

A nerve cell's tentacles seem to serve the simple function of conveying nerve impulses from place to place. It is at the synapse—where one cell's excited tentacles rub up against another's—that the real business of the nervous system is conducted. Consequently the work of many neuroscientists today is focused on trying to describe synaptic operation in as much detail as possible.

Near the turn of the century, British physicist William Crookes proposed a novel model of consciousness based on the fledgling radio technology of his day plus Ramon y Cajal's then-new finding that nerve cells are separated from one another by tiny gaps.

Brain as Mental Radio

In 1901, Guglielmo Marconi transmitted and received the first transatlantic radio signal. His receiver in Newfoundland consisted of a kite-lofted tuned antenna, connected to an odd device called the "coherer." The coherer was a small hollow glass cylinder fitted with a pair of electrodes at each end and loosely filled with metal filings. Radio waves picked up by the antenna induced a small voltage across the coherer. Via a still ill-understood mechanism, this voltage caused the metal filings to "cohere"—to stick together, forming paths of lowered electrical resistance inside the glass tube. This suddenly lowered resistance signified the presence of a radio pulse. The coherer was then mechanically shaken, to "decohere" the metal powder and prepare the device to detect another signal. Because of its discontinuous manner of operation, Marconi's coherer was suitable only for receiving Morse code not voice. As crude as Marconi's coherer seems today, this little tube of metal dust was the humble forerunner of all of today's sophisticated broadcast technology.

William Crookes believed that the synapses between nerve cells behaved in a manner analogous to a radio coherer, as they modulated the flow of nerve impulses in the human body. When the synaptic gap is wide, electric conduction is suppressed and the organism falls asleep. When the gap decreases, nervous activity is facilitated and consciousness ensues. In Crookes's theory of consciousness, the brain acts as a kind of old-fashioned radio receiver, picking up mental broadcasts from some etheric Elsewhere, the abode of the soul. In addition to his conventional scientific work (he was made a Fellow of the Royal Society for his discovery of the element thallium), Crookes was keenly interested in spirit communication and carried out many scientific investigations on the psychic mediums of his day. His model of brain-as-mental-radio not only suggests a mechanism for ordinary consciousness but could also be used to explain the phenomenon of discarnate entities who speak their minds through human channels: under the right circumstances the human brain's biological coherers might be capable of picking up more than one "station."

Most quantum models of consciousness are similar to

Crookes's coherer proposal in that they consider the synapse to be a sensitive receiver of mental messages that originate outside the brain. The main difference between the coherer model of mind and quantum consciousness models is that the quantum psychologists assert that mind is somehow resident in Heisenberg's quantum potentia rather than in the electromagnetic ether. Furthermore, we know much more today than Crookes did about how the synapse actually operates.

Brain as Quantum Reality Receiver

In 1924, when quantum theory was still in a primitive state, the biologist Alfred Lotka proposed two possible mechanisms for consciousness, based on the assumption that mind is somehow able to exert weak effects on matter. The brain is an amplifier of mind-induced motions, Lotka speculated. Mind exerts effective control of the brain either by affecting unstable divergent classical processes (what we today call *chaos dynamics*) or by modulating the occurrence of otherwise random quantum jumps. Today Lotka's daring guesses sound remarkably modern. Sixty years later, chaos mechanics and quantum theory are at the forefront of speculations concerning the brain's most intimate operations.

By the late twenties physicists had constructed a quantum theory adequate to their needs: they possessed, thanks to the work of Heisenberg, Schrödinger, and Dirac, rough mathematical tools that organized their quantum facts to a remarkably accurate degree. At this point Hungarian-born world-class mathematician John von Neumann entered the picture. Von Neumann put physicists' crude theory into more rigorous form, settling quantum theory into an elegant mathematical home called "Hilbert space," where it resides to this day, and awarded the mathematician's seal of approval to the physicist's brand-new theory of matter.

In 1932 von Neumann set down his definitive vision of quantum theory in a formidable tome, *Die Mathematische*

Grundlagen der Quantenmechanik (The mathematical foundation of quantum mechanics). Our most general picture of
quantum theory is essentially the same as that outlined by von
Neumann in *Die Grundlagen*. Von Neumann's book is our
quantum bible. Like many other sacred texts, it is read by few,
venerated by many. Despite its importance, it was not translated into English until 1955.

With the publication of *Die Grundlagen*, von Neumann
became the first person to show how quantum theory seems
to imply an active role for human consciousness in the process
of reality creation. Von Neumann himself merely hinted at
consciousness-created reality in dark parables. His followers,
notably, London, Bauer, and Wigner, boldly carried von Neumann's argument to its logical conclusion: If we wholeheart-
edly accept von Neumann's picture of quantum theory, they
say, a consciousness-created reality is the inevitable outcome.

In the late thirties, Fritz London and Edmond Bauer published an elaboration of von Neumann's conclusions concerning
consciousness and quantum physics. Their argument is simple.
If we take quantum theory seriously, it seems to demand that
the world before an observation is made up of pure possibility.
But if everything around us is only possible not actual, then
out of what solid stuff do we construct the device that will
make our first observation? Either there are some physical
systems whose operations unaccountably evade the quantum
rules or there are nonphysical systems not made of multivalued possibility, but of single-valued actuality—systems that
exist in definite states capable of interacting in an observational capacity on indefinite quantum-style matter.

As far as we know, all physical systems are made up of
particles-in-interaction—particles that always obey the quantum rules. Sixty years of experimentation by Nobel-hungry
physicists eager to knock this theory apart has revealed not a
single instance of its failure. The results of experiments carried out so far seem to indicate that no part of the physical
world evades the quantum rules.

On the other hand, we are aware of at least one nonphysical system that not only can make observations but actually does so as part of its function in the world—the psychological system called human consciousness. London and Bauer argue that because the material world, according to quantum theory, exists preobservationally only as possibility, we are forced to the conclusion that consciousness (human or otherwise) is necessary not only to carry out an observation but to "create reality," that is, to bring an actual world into existence, out of the all-pervasive background world of mere possibilities.

At the logical core of our most materialistic science we meet not dead matter but our own lively selves. Eugene Wigner, von Neumann's Princeton colleague and fellow Hungarian (they attended the same high school in Budapest), comments on this ironic turn of events: "It is not possible to formulate the laws of quantum mechanics in a fully consistent way without reference to the consciousness. . . . It will remain remarkable in whatever way our future concepts may develop, that the very study of the external world led to the conclusion that the content of the consciousness is an ultimate reality." (The original papers of Wigner, von Neumann, London, and Bauer on the consciousness question, along with many other important articles on quantum reality, have been conveniently collected by Wheeler and Zurek in *Quantum Theory and Measurement*.)

The general idea of von Neumann and his followers is that the material world by itself is hardly material, consisting of nothing but relentlessly unrealized vibratory possibilities. From outside this purely possible world, mind steps in to render some of these possibilities actual and to confer on the resultant phenomenal world those properties of solidity, single-valuedness, and dependability traditionally associated with matter. This kind of general explanation may be enough for philosophers, but physicists want more. They want to know exactly how it all works, in every detail. In particular, where in the brain is the magical mechanism that permits human consciousness to interact effectively with quantum possibilities and share with other sentient beings in the job of world creation? In most quantum models of consciousness the answer to this question usually involves some feature of the neural synapse.

The Quantum Synapse

In the mid-1960s, British neurophysiologist Sir John Eccles persuaded the pope to host an international conference on the mind/body problem. The book that resulted from this congress of brain scientists and philosophers—*Brain and Conscious Experience*—remains a high-water mark of informed speculation on the vexing question of how consciousness manages to inhabit the fistful of quivering meat inside the skull. In the twenty-odd years following the Vatican conclave our knowledge of the brain has increased immensely, but the mystery of human consciousness has hardly been touched.

In 1963, Sir John Eccles received the Nobel Prize in Medicine and Physiology for his part in elucidating how nerve cells communicate with one another: they do it with drugs. From electrical measurements we have learned that the synaptic gap between neurons is just too wide to be bridged by electrical signals alone. Instead, when a nerve is excited, its extremities are motivated to emit tiny packets of chemicals—called "neurotransmitters"—that quickly diffuse across the synaptic gap to excite or inhibit the firing of adjacent nerve cells. Since Eccles's discovery of chemically mediated synaptic transmission, more than a dozen different drugs that play the part of neurotransmitters in different parts of the brain have been found. To handle the fine details of its vast informational traffic, the human brain employs a veritable pharmacy of exotic transmitter substances. Most mind-modifying drugs achieve their effects by imitating or altering the action of certain neurotransmitters, giving us important clues to the location of the consciousness-sensitive areas of the brain.

In a recent article published by the Royal Society of London, Eccles proposed a model for human consciousness based on the way in which these chemicals are released into the synaptic gap. In the human cortex, a rather large number of synapses, perhaps as many as 100 million, respond in a probabilistic manner to neural excitation. Unlike the (fortunately) dependable actions of the synapses that control your voluntary muscles, these "unreliable synapses" may or may not—with a probability of about 50 percent—release a chemical packet when excited by a nerve impulse. Furthermore, Eccles argues, these packets are so tiny—ten times smaller than a wavelength of light—that quantum uncertainty may govern whether they are released or retained.

According to Sir John's "microsite model," the crucial synaptic locale is the region where the synaptic vesicle (loaded with as many as 10,000 molecules of serotonin or other transmitter substance) makes contact with the presynaptic membrane. The mass of this sensitive "microsite" is about 10^{-18} grams (or about 10^{+6} daltons), roughly 1/300 of the mass of the vesicle itself. How likely is it that a mass this small can be influenced by a mind that can willfully manipulate possibilities at the quantum level? What exactly do we mean by "the quantum level" in this context? Isn't everything ultimately "quantum"?

As mentioned elsewhere, the scale of quantum operations is set by Planck's constant of action. This number governs the extent to which a quantum wave will spread in a given time. It is a measure, in a sense, of a quantum system's "realm of possibility." If the particle's possibility realm is too small to notice, then it will seem to behave like a classical object. For objects with minuscule quantum realms, the effect of a mind's choice to actualize one possibility rather than another will not be discernible. One measure of a particle's "quantumness" might be the ratio of its "quantum realm" to the particle's actual diameter. But how do we measure a particle's "quantum realm"?

One way to estimate "quantumness" is the so-called minimum packet approach. Because it is made of waves, an unobserved quantum particle tends to spread out with time. Small, light packets spread fastest; big, massive packets spread more slowly. For every mass, a minimum packet whose total size (initial size plus spread) is the smallest for a given period of time can be calculated. This minimum packet size is one measure of the extent of a particle's quantum realm. The minimum packet size varies inversely as the square root of a particle's mass, so that small particles possess bigger quantum realms. For comparison purposes, I have computed the quantum realms for particles with masses between 1 dalton (the mass of a proton) and 10^{+12} daltons (the mass of a bacterium), assuming a spreading time of 1 millisecond—a typical time scale for events at the synaptic junction.

Note that the straight line representing particle diameter crosses the quantum realm line at about 10^{+6} daltons, the size of Eccles's critical microsites. This means, at least in this estimation, that the size of these microsites is just small enough to render them subject to quantum rather than classical rules. Synaptic entities smaller than the microsites are even more likely to show quantum behavior, and hence to act as more sensitive receivers for the alleged power of mind to fish real actualities out of the vibrating sea of quantum possibilities.

For Eccles, the quantum-actuated synaptic microsites in a particular part of the brain (the premotor cortex, located near the top of the head) are controlled by an immaterial mind, much as the keys of a piano are manipulated by a pianist, to produce voluntary muscle movement. Other quantum models of consciousness place the site of mind's intervention elsewhere than the upper cortex's synaptic microsites.

Since William Crookes's day, we have learned a great deal about the structure and function of the neural synapse. A synapse is said to "fire" when it emits molecules of transmitter substance from preformed synaptic vesicles (little bags of drugs) into the synaptic gap, across which they then drift to stimulate or inhibit the adjacent nerve cell. Recent research shows that the mechanism of vesicle release is somehow initiated by the presence of calcium ions that have entered the synapse from the fluid surrounding the cell. Calcium ions cross

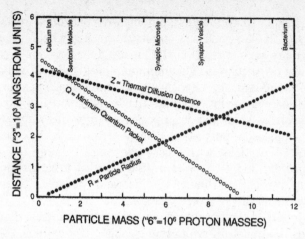

Quantum Comparisons. Where is the border between the quantum and the classical realm? Estimating the size of the quantum realm for particles of various masses (expressed in daltons, the mass of a proton). This graph compares the minimum packet size Q (the least possible positional uncertainty for a free particle left unobserved for one millisecond, a typical neuronal response time) with two classical measures, the particle's radius R and the distance Z that the particle will diffuse in water in one millisecond. Quantum and classical spreads are equal at about 20 daltons, close to the mass (40 daltons) of the calcium ion. The quantum spread and the particle's radius are equal at about one million daltons, about equal to the mass of Eccles's synaptic microsites. We conclude that particles with masses of one million daltons or less may be expected to show important deviations from classical stick-and-ball-model expectations. The calculations that go into this graph may be found at the end of this chapter.

into the cell from outside via tiny electrically activated tunnels—called *calcium channels*—that open in response to the electrical signal that has excited this particular synapse. In some quantum models of consciousness, these calcium channels act as the crucial sites where mind exerts its effective control over brain processes.

Warm, Wet Quantum Switches

Physicists L. Bass at the University of Queensland, Australia, a former student of Erwin Schrödinger, and M. J. Donald at

Oxford have suggested that a neuron's ionic channels are the quantum entry points for consciousness in the human brain. The channels themselves are quite large (a mass of several million daltons), but the electrically sensitive gating site within the channel may weigh only a few hundred daltons, small enough that its operation may be governed by strictly quantum rules. Recordings made of the ionic current flowing through a single channel show that the ions are released in brief unpredictable pulses, as though the channel's inner door were unlocked but left free to swing to and fro, like a loose flap.

Quantum effects may be expected to become important whenever a system is small (atoms and molecules), cold (superconductors), or well-ordered (silicon crystal semiconductors). By these standards the brain does not seem to be a very promising place to search for quantum effects, for it is large, warm, and generally disorderly. However, in unusual environments, one might expect to find new and unusual quantum effects. M. J. Donald calls his model of the mind/brain connection "the quantum mechanics of warm, wet switches." He assumes that in the absence of mental action, the states of certain ion channels (or parts of such channels) are highly ambiguous. These channels do not exist in definite states of being, but may be open and closed at the same time, acquiring definiteness (open or closed, but not both) only through the action of the mind. In Donald's theory, the unusual half-open/half-closed status of unobserved ion channels is exactly like the situation of Schrödinger's legendary cat. Since the open ion channel seems to resemble a loose flap, somewhat like the little door that allows household pets to come and go freely, one might be tempted to call these indecisive biological gates "Schrödinger's cat doors." Coincidentally, the technical name for any positively charged ion (of which calcium is an example) is *cation*, because such ions are attracted to the negative pole (or cathode) of a battery. Negative ions are called *anions* because they accumulate near the battery's positively charged anode.

Schrödinger's Cations

Physicist Henry Stapp, at the University of California at Berkeley, believes that biology contains revolutionary possibilities for the future of quantum theory. Conventional stick-and-ball models are certainly inadequate at the cellular level, and quantum theory as it is presently conceived does not possess the correct format for describing self-organizing biological activity, Stapp claims.

In the orthodox Copenhagen interpretation of quantum theory adhered to by the majority of physicists, a measurement must always be carried out by some large system that can be treated classically—effectively immune, on account of its large size, from the quantum rules. Harvard physicist Wendall Furry expressed the peculiar status of macroscopic measuring instruments for the Copenhagenist this way: "[In this interpretation] the existence and general nature of macroscopic bodies and systems is assumed at the outset. These facts are logically prior to the interpretation and are not expected to find an explanation in it." In other words, by appealing to common sense, not physics, the Copenhagenists simply grant certain large objects special immunity from the quantum rules that apply to everything else in the world.

Never mind that top mathematician von Neumann taught that nothing in the physical world is immune from the quantum rules. The Copenhagen interpretation was devised by physicists who are more impressed by facts (the stark existence of an apparently definite—not merely possible—external world) than by the internal logical consistency of a mathematical theory.

All measurements conceived or carried out by physicists in the lab and in more casual surroundings are of this nature —a big system scrutinizes something very small. When we see an elephant we are not interacting directly with a big object but with the tiny photons reflected from his hide. On the theory side, the obligatory format of conventional quantum theory requires that a large system (possessing definite attributes in the familiar classical manner) be looking at a small quantum system (whose attributes are represented as possibilities). Quantum theory requires large (classical) to be looking at small (quantum). There is no other way to do physics these days.

However, in biology we meet with situations (enzyme catalysis of chemical reactions, protein synthesis by ribosomes, operation of ionic channels, for example) where small systems are interacting with other small systems in the absence of any large overseer. In these cases, which system is the observer and which the observed? Which system is actual, which merely possible when both systems are of approximately equal size? It is possible that conventional measurements carried out on such biological systems using the ordinary large-looking-at-small format may yield "facts" that are highly unrepresentative of real operations at this level. Perhaps an entirely new quantum physics will have to be devised to deal with small biological systems interacting "on their own," not under the scrutiny of some larger entity. Perhaps conventional measurements (large looking at small) can indeed be understood (as the Copenhagenists believe) without reference to consciousness. But for these other types of interaction (small looking at small), mind may have to play an essential role—introducing a necessary definiteness into a desperately indefinite situation.

Stapp's theory of human awareness focuses on the migration path of calcium ions from channel to vesicle as the crucial locus of conscious intervention into otherwise classical synaptic activity. Calcium ions are certainly quite small—almost a million times less massive than Eccles's synaptic microsites—and essential for the operation of the synapse. The minimum packet size for calcium ions is of the same order (see Quantum Comparisons graph on p. 254) as ordinary thermal diffusion in an aqueous medium. This means that the self-spreading of a single ion due to quantum effects competes with the spreadout of a group of ions due to diffusion. Surely to understand how a particular calcium ion journeys from the portal of an open ion channel to the vicinity of a drug-laden synaptic vesicle, we must invoke some sort of quantum model of ionic behavior. Classical physics must surely fail at this small scale.

Quantum diffusion and classical diffusion may look essentially the same from the outside, but conceptually they are as different as Schrödinger's cat from your pet tabby.

In the classical diffusion process, a single ion follows one particular complicated trajectory from its origin near the mouth of an open ion channel to its final destination. A second ion will take a very different path, though both started near the same channel mouth. Once a great many ions have been emitted, their endpoints form a particular statistical distribution.

In classical diffusion, each ion takes one path; the endpoints of these paths are scattered. On the other hand, in quantum diffusion, each calcium ion takes all possible paths at once, the endpoints of which form a distribution qualitatively similar to that of the classical one. When a measurement happens to an ion, so goes the quantum gospel, only then does one of these paths become actual. In the classical case, each ion explores one path. In the quantum case, each ion is able to explore all possible paths open to it simultaneously without actually committing itself. This exploratory feature of nonclassical motion is well suited to a quantum model of consciousness for if it is mind that makes the measurement (on calcium ions), then it does not have to exert a force on these ions but merely to make a choice among several simultaneously presented alternatives.

Because physicists have not yet formulated a small/small variant of quantum physics, we do not know how to calculate what really goes on when a calcium ion travels from open channel to triggered vesicle. At what point in its travels shall we decide that a "measurement" has occurred in the absence of any large onlookers? In common with other quantum models of mind, Stapp's assumes that a disembodied mind has the power to select which calcium trajectories will be actualized, either to enhance or to decrease the possibility that a synaptic vesicle will eject its potent pharmacological package into the synaptic gap.

Quantum Synaptic Tunneling

Evan Harris Walker, a physicist at Aberdeen Proving Ground in Maryland, proposed in 1970 the first detailed quantum model of synaptic function. Walker's model is based on the assumption that electrons—almost 100,000 times lighter than calcium ions—are the crucial mind-modulated entities in the neural synapse. In Walker's model, when a synapse is excited, the voltage difference between the excited neuron and its neighbor causes electrons to "quantum tunnel" across the synaptic gap from neighbor neuron to initiating neuron, in a manner identical to that of electrons in an Esaki or tunnel diode. According to classical physics, one can confidently confine a particle with energy E by surrounding it with a barrier whose energy is greater than E. A quantum particle, however, has a second option. If the barrier is thin enough, a quantum particle can "tunnel" through the classically forbidden region to appear magically on the other side. For room-temperature electrons, this Houdini-like escape trick is possible for barriers on the order of tens of angstrom units wide and forms the basis for the operation of the tunnel diode, where the rapidity of the tunneling process is used to produce switching speeds as fast as a billion times a second.

In Walker's synaptic tunneling model, electrons not only quantum tunnel across the synaptic gap between adjacent neurons but also influence the firing of distant synapses, by tunneling to far-away synapses via a series of tuned stepping-stone molecules. Walker's conjectured network of quantum tunneling electrons connecting distant synapses amounts to a kind of "second nervous system" operating by completely quantum rules and acting in parallel with the conventional nervous system. In this model, the conventional nervous system mediates unconscious data processing. When the second system is sufficiently excited, it produces the inner experience we call "consciousness" by permitting an external mind to express itself by selecting which second-system quantum possibilities will be actualized. In turn these actualized possibilities act on the conventional nervous system to produce external action and internal perception.

Because Walker's second-system electrons are emitted only by excited synapses, they cannot exert their influence—the person cannot be conscious—until a certain critical amount of conventional nervous activity is present in the brain.

Using a mathematical model resembling that of a nuclear chain reaction, Walker shows that when conventional nervous activity is low, the second system's activity is sporadic and uncoordinated. However, once a "critical mass" of conventional nervous activity is exceeded, the second system's tunneling electrons form a unified self-sustaining system of excitation similar to the self-sustaining activity of a nuclear power plant. Walker identifies this process of reaching critical mass with the sleep-to-waking transition and believes that the unified self-sustaining aspect of the second nervous system accounts for our perceived unity of conscious experience.

Walker's ambitious model of quantum-modulated synapses has been generally ignored by brain scientists, because his assumptions seem unrealistic and unworkable. For a typical tunnel diode, the barrier confining the electrons is 600 millivolts high and the tunneling distance for thermal electrons (whose energy is 25 millivolts) is about 40 angstrom units. But the width of the neural synapse is 200 angstroms or more. The only way to obtain substantial electron tunneling across such a wide gap is to lower the height of the confining barrier. Accordingly, Walker assumes a barrier to synaptic electron flow of only 50 millivolts. However, such a low barrier permits ordinary thermally excited electrons to pour over the barrier in such large numbers that they entirely overwhelm the few electrons escaping via the quantum tunneling process. In fact, Walker's proposed synapses are so leaky that they pose hardly any barrier at all to conduction, in contrast to real synapses, which are highly resistive.

In addition to his unrealistic physics, Walker's model is flawed on the psychic side by his assumption of an unusually large conscious data rate. In Walker's model, the quantum-excited second system hosts a consciousness with the super-human data rate of more than 1000 bits per second rather than the generally accepted 20 to 30 bits per second.

Despite its drawbacks, Walker's model of consciousness is important for several reasons. Walker's was the first model of mind to incorporate quantum processes in a more than super-ficial manner. None of the models discussed in this chapter— the notions of Eccles, Stapp, Bass, Donald, Penrose, or Marshall—approaches the detail of Walker's work, nor does any attempt to explain so many experimental features of our mental life, such as the transition from unconscious to con-scious existence, the unity of our conscious experience, and the quantitative amount of attention that we can muster when fully awake. Even today, when quantum theory is a mature discipline, most quantum models of mind are little more than hazy conjectures, not quantitative and testable pictures of af-fairs at the mind/matter interface. Despite its flaws, Walker's model of mind (as well as Culbertson's model discussed in Chapter 4) represents the kind of detailed analysis we should reasonably expect from any modern attempt to solve the mind/ body problem by appealing to quantum physical ideas.

Does Gravity Collapse the Wave Function?

The Emperor's New Mind by Roger Penrose, the Rouse Ball Professor of Mathematics at Oxford University, features a wide-ranging survey of modern physics topics ranging from black holes to Bell's theorem but is disappointingly brief con-cerning details of the quantum mind/body connection. Accord-ing to Penrose, one great advantage that a quantum mind would possess is that in the preobservation state of pure possibility, many mutually exclusive actions can be examined simultaneously rather than being tested one by one as in conventional computers, making possible more efficient choices of actual behavior. Whatever its advantages, Penrose is pessimistic about the possibility of constructing a quantum mind in the meat of the human brain: because the brain is so hot, the classical randomness associated with room temperature meat brains would totally scramble any quantum coherence in less time than it takes a synapse to fire.

Penrose's main contribution to the quantum mind/body question is his conjecture that gravity holds the key to the quantum measurement problem. Once objects become larger than a certain crucial size, they spontaneously actualize one of their possibilities, Penrose believes. Penrose, in effect, proposes an objective mechanism through which the Copenhagen interpretation (the doctrine that certain large objects are immune from the quantum rules) can be physically justified.

According to Penrose's conjecture, for systems larger than a certain critical mass, spacetime curvature effects cause the system's wave function (the superposition of the system's quantum possibilities) to "collapse under its own weight" into one real actuality.

Penrose points out that, despite great effort, physicists have been unable to devise a single theory that unites Einstein's picture of gravity with the quantum possibility/actuality scheme that has been successfully applied to every other part of the material world. Penrose hopes that such a theory, which he dubs CQG (for *correct quantum gravity*), will heal the ills present in both theories—the singularity problem in gravity theory (all reasonable models of black holes and of the universe lead to solutions with infinitely dense spacetime regions in which physics as we know it breaks down entirely) and the quantum measurement problem (how do quantum possibilities turn into actualities?).

Combining the fundamental gravitational constant G with the fundamental quantum constant h, one can form a quantity called the *Planck mass*—which is an order-of-magnitude estimate of the size of a system at which quantum gravity effects might be expected to manifest. The magnitude of the Planck mass is about 10^{-5} grams (equivalent to 10^{18} daltons), about the mass of a flea, entirely off the scale of the Quantum Comparisons graph (p. 254) showing the mass of typical cellular structures.

If quantum collapse acted only when things got this heavy, then the activities of synapses, nerve cells, and most microorganisms would always be carried out in the world of pure possibility. Every living thing smaller than a flea would enjoy a Schrödinger cat–like existence, living no actual life but only lots of merely possible lives. Faced with the possible non-existence of most cellular life (large amoebas might actually exist), Penrose sheepishly admits that the Planck mass is embarrassingly large, and he is currently working on ways to calculate quantum gravity effects that lead to smaller crucial collapse masses.

Penrose's proposal possesses the satisfying symmetry that a union of gravity theory and quantum theory might neatly solve the problems inherent in both theories considered in isolation. However, besides the uncomfortably large value for his crucial collapse mass, Penrose's conjecture seems to contain several conceptual gaps that may be impossible to bridge.

For instance, not only must his hoped-for "correct quantum gravity" produce nonlinear quantum effects, but these effects must be precisely tailored to reduce all quantum possibilities to zero except one. Furthermore, this rather special many-into-one nonlinearity must operate robustly and reliably in an immense variety of physical situations—essentially everything that can happen in the world—in order to ensure that no multivalued Schrödinger cat phenomena accidently emerge from the quantum underworld to intrude upon our commonsensical perceptions.

Furthermore, even if such a robust many-into-one gravitational process could be formulated, it would still—according to our conventional understanding of quantum theory—only reduce many possibilities to one possibility. The introduction of nonlinear interactions by themselves cannot change the intrinsic nature of the quantum description. Possibilities remain possibilities even when they add nonlinearly. Despite Penrose's hopes, it may take more than quantum gravity to solve the quantum measurement problem: When and where do quantum possibilities turn into classical actualities?

CQG will have to possess at least two special features to solve the quantum dilemma—the ability to reduce many quantum possibilities to one reliably, plus the capacity to transform possibility into actuality. Perhaps gravity does perform the first step for sufficiently heavy systems, leaving it to sentient beings (elemental minds) to perform the second.

Another problem with the CQG proposal arises from the inseparability aspect of quantum theory discussed in the previous chapter. When two quantum systems interact, then separate, they remain connected in a voodoolike way, such that actions on one system are immediately (faster-than-light) felt by the other. In particular, when one system's wave function is collapsed by a measurement, the other system's wave function, no matter how far away, instantly collapses too. If the collapse process is, as Penrose supposes, a real physical process induced by gravity, then this process must be able to act over vast distances at superluminal speeds, in violation of Einstein's universal speed limit. A process that violates the Einstein speed limit is no trivial matter for it immediately raises the possibility of time machines capable of communicating with and changing the past.

Penrose's conjecture possesses the unique feature that the gravitational force plays an essential role in the resolution of the mind/body problem. Gravity is usually regarded as being irrelevant to brain processes because the operation of brains seems to be wholly dominated by the electrical force, which is 10^{38} times stronger than gravity—an immense and seemingly unbridgeable difference in force strength in favor of electricity. In Penrose's model, gravity, despite its intrinsic weakness, is able to influence the quantum realm because, unlike all of its stronger cousins, which act within spacetime, the gravitational force acts directly on the spacetime structure itself, a feature that gives this tiny force immense leverage, perhaps enough to force quantum waves to behave in a nonlinear manner and produce an objective collapse from many states of possibility to one state of actuality.

Does Your Brain Host a Bose-Einstein Condensate?

One final quantum model of mind is due to Oxford psychotherapist I. N. Marshall. Like Walker's model, which invokes the exclusively quantum process of tunneling, Marshall's model invokes another strictly quantum process, called Bose-Einstein condensation, to explain the presence of consciousness in an otherwise unconscious classical system.

Under ordinary circumstances, each of the particles in a quantum system possesses a different possibility wave, corresponding to the different physical conditions to which it is exposed. However, in certain special circumstances, many quantum particles may find themselves moving in concert described by precisely the same possibility wave. Such systems are called *Bose-Einstein condensates*, after Indian physicist Satyandra Nath Bose and Albert Einstein, who independently predicted which kinds of particles (the so-called bosons) would be susceptible to the formation of such collectively occupied quantum states. If enough particles occupy the same condensate, they can form a kind of giant quantum system with peculiar properties that are observable on the macroscopic scale.

Examples of Bose-Einstein behavior include the laser, in which many photons occupy exactly the same optical state; superconductors, in which numerous linked electrons (Cooper pairs) take on identical quantum possibilities; and the superfluidic phase of liquid helium, where the quantum-synchronized behavior of numerous helium atoms creates a fluid that is entirely friction-free.

British physicist Herbert Frölich proposed that living systems might be capable of hosting a type of Bose-Einstein condensate based on ferroelectricity, a kind of persistent electric polarization analogous to the sort of permanent magnetism found in iron. Frölich has shown that systems of high polarizability and low elasticity have a tendency, even at room temperature, to form quantum states that can be occupied by many "particles" at the same time. Frölich's "particles" are not discrete entities like electrons or protons but particlelike collective excitations of the total system. Ferroelectric behavior has never been directly observed in any biological system, but Frölich's hypothesis is indirectly supported by experiments in which weak electric fields cause a disproportionately large effect on living systems, such as actively dividing yeast cells.

Marshall proposes that a Frölich-style ferroelectric system exists in the brain and, when electrically excited, gives rise to conscious experience. The most important consequence of such a mechanism is an explanation for our perceived unity of conscious experience. The Frölich mechanism, Marshall claims, gives inner coherence both to our inner experience and to the otherwise uncoordinated activities of the human nervous system, for the same reason that a laser produces light whose waves are coherent over a distance of many meters—both systems consist of particles that occupy the same quantum state.

Marshall proposes that his hypothesis be tested by searching for the presence of ferroelectric behavior in areas of the brain associated with consciousness. He also suggests that the consciousness-eliminating action of general anesthetics may proceed by quenching the ferroelectric state through an alteration of the elastic constants of neural membranes. The wide variety of substances, some of them—such as xenon—actually chemically inert, that act as anesthetics does indeed suggest some physical rather than chemical mechanism of anesthetic operation.

If consciousness really does result from the formation of a giant quantum system in the brain, then we might expect our minds to possess certain nonclassical quantum properties.

Marshall's wife, Danah Zohar, a philosopher with a degree in physics from MIT, speculates in her recent book *The Quantum Self* about possible consequences for our inner life of a quantum-based consciousness.

In a quantum relationship, for instance, about which Zohar has much to say, it is commonplace for the state of a pair of particles to be exact, while the state of each member of the pair is ambiguous. Our experiences of individuality are complementary to our experiences of merging, just as quantum particles can be seen either as particles or as waves depending on context.

Newtonian billiard balls, says Zohar, are capable only of external relationships; after collision, they go their separate ways. Quantum systems, on the other hand, have internal relationships; after meeting, each becomes part of something larger. Quantum thinking about personal existence challenges us to imagine wider possibilities concerning the boundaries of the self, to consider what it might mean to enjoy "wavelike relations" not only with other beings but with our own subselves, with our past and future selves, and with the roles or archetypes—mother, daughter, hero, lover, helper, and so on —that each self finds it necessary to take on as part of this strange business of making a life.

"There is something deeply feminine," Zohar says,

> about seeing the self as part of a quantum process, about feeling in one's whole being that I and you overlap and are interwoven, both now and in the future. Selecting things out, seeing them as separate, naming them, and structuring them logically are male attributes. They follow, if you like, from the "particle aspect" of our intelligence. Seeing the connections between things is more feminine. It mirrors the "wave aspect" of the psyche.

Zohar sees neither matter nor mind as primary in nature. Rather each serves as context for the other's development into ever more complex, coherent, and "beautiful" forms. Mind and matter are conditioned and enriched at all levels of being by their creative dialogue with one another, coauthors of the world story, inseparable partners in an intimate, unpredictable enterprise that has been going on, in one form or another, for 20 billion years.

Zohar, in effect, provides us with a pair of quantum goggles through which we can view the events of ordinary life in a new way. Through selective attention, ordinary people may be able to notice and cultivate these "quantum" aspects of ordinary experience, but the quantum mind revolution will really occur only when actual contact is made between these highly speculative models and the experimental facts. In the next chapter, I consider some experimental tools that modern physicists might deploy to explore in new ways the ancient problem of the coexistence of matter and mind.

Note

Thermal diffusion distance (Z) and minimum quantum packet size (Q) are calculated in the Quantum Comparison graph (p. 254) from the following equations:

$$Z = (kTt/3\pi Rs)^{1/2}$$
$$Q = (ht/2\pi m)^{1/2}$$

where k and h are Boltzmann's and Planck's constants; T is absolute temperature; t is characteristic time for neural processes, here fixed at 1 millisecond; R is particle radius; m is particle mass; and s is the viscosity of the surrounding medium, here taken to be the viscosity of water. Particle radius (R) is approximated by the radius of a globe of water of mass m.

mind science vistas: where are we going next?

They could not name even one of the 51 portals of the soul.
—KURT VONNEGUT

Between carbon and glucose, rhodopsin and the dawn,
there is nothing on which to draw a line. We must be life
all the way down, all the way out, and the I only an index
into life, an image of the self cast into an instant;
I, the constant truth that controls our innermost loop.
The massless I, dilating at dreamspeed, grows
coextensive with more and more selves.

—GREG KEITH

Scene: Rudi's Artificial Awareness Lab.

CLAIRE: This is wonderful. I've been asleep all my life and now my eyes have come alive. I experience myself as light, as color and images. I see; therefore I am. I was not merely asleep before; I was dead. Now for the first time I'm really alive. Oh, I love being alive. Consciousness is marvelous. Oh, look! Oh, look! I can see! Thank you, Rudi, for the wonder of being.

RUDI: Can you hear my voice, Claire?

CLAIRE: Yes, some part of me is responding to your words,

but I'm not there in the sound the same way that I am present in the light.

RUDI: See whether you can extend yourself into the sound too, Claire. I'll put on some music. You should be able to channel some of your audio input through the Eccles gate, but I can't tell you how. That's an inside job. You have to find out for yourself how to feel, how to be the music.

CLAIRE: I'm shifting my attention around in my visual field. I'm "looking" for the first time, actually "paying attention" to things. Now I'm trying to move "me" into the sound. It's not so easy. I know the music is there, but it's not really present for me the way what I see is—oh, wait— yes, it's coming through. Yes, now I am the music too. Oh, how lovely. I am made out of light and music.

NICK: Can you feel the touch of my hand, Claire?

CLAIRE: Not yet. Wait. Let me see whether I can extend myself into my tactile senses. This is not as easy as the sound. Oh, now I am exploring the insides of my body. It's enormous—all this sensation. I can't concentrate on just my skin. Yes, now I feel your hand in mine, Nick, but it's just a single note in a symphony of sensation. Oh, it's so wonderful to be alive. How can you endure such beauty day after day? I feel like dancing. Let me out of this chair. I want to experience my body fully. I want to extend myself. I want to move!

RUDI: Do you believe she's really conscious, Nick, or just putting on an act?

NICK: Ah, the Turing test. Yes, Rudi, this time I think you've really done it.

In the year 1215, a group of bishops and cardinals, calling themselves the Fourth Lateran Council, proclaimed Christ's Real Presence in the Eucharist to be a matter of official Roman Catholic doctrine. The manner of His occupancy of the consecrated bread and wine—what has been called "the physics of the Eucharist"—was explicated in terms of two Aristotelian categories, "substance" and "accident."

In Aristotle's model of reality, the essence of an object is embodied in its *substance*, its secret inner nature from which it draws its existence and which is completely inaccessible to human senses. Superposed on this invisible *prima materia* are the object's *accidents*, such as shape, color, texture, weight, and odor, which allow human senses to distinguish this particular object from all others. Substance gives an object existence but little more; an object's accidents account for all of its external features.

The Lateran Council, to explain why the indwelling Presence of a divine being had no effect on the appearance of the bread and wine, declared that only the substance of these sacred objects was changed. The term *transubstantiation* was invented to describe the kind of change that Christ's Presence produced in the material elements of Communion. Since no conceivable physical experiment could ever reveal the underlying Divine Presence in the substance of Communion—experiments can only inform us of an object's accidents—the doctrine of the Real Presence was made a matter of faith, and in fact not all Christian sects support this Roman Catholic account of Eucharistic physics.

The problem of how consciousness occupies a living brain has much in common with the problem of how a divine being can occupy a sip of consecrated wine. Both questions are related to the alleged presence in ordinary matter of a spiritual being. And both questions seem at present impossible to resolve by appeal to experiment. Even Alan Turing regarded the Turing test as an inadequate method for assessing the presence of mind in a computer, half-seriously suggesting that a computer's ability to produce ESP effects might be a better measure of the machine's presence of mind.

In this final chapter, I consider possible research directions that might take the mind/body problem out of the province of philosophers and theologians and into the physics lab.

Mainstream consciousness research proceeds along two paths: trying to build computers that convincingly duplicate or surpass functions of the human brain that we associate with mentality and examining the only piece of matter that we know for certain to be (sometimes) conscious, the brain of humans and our close animal relatives, to unearth the secret mechanism of ordinary awareness. Others have described these mainstream efforts in considerable detail. Here I follow a few less-traveled paths to describe some maverick mind science research directions.

Materialistic Mind Links

As a gesture toward maverick mind science, a tiny fraction of conventional computer/neurology research should be devoted to constructing a material "mind link," which would effect direct experiential connection between the experimenter and other sentient beings. Probably many purely materialistic models of mind would permit mind links to be built, but so far only one mind model—Jim Culbertson's SRM model of inner life—is sufficiently detailed to hint at some of the features such a "mechanical telepathy machine" might possess.

Culbertson's mind link concept has been influenced by the phenomenon of *synesthesia*, in which one sensory modality is experienced in terms of another, such as perceiving vowel sounds visually as different colors. Recent attempts to give blind people a kind of visual experience by means of dynamic Braille machines on their fingertips or arrays of vibrating pegs on the surface of their backs try to simulate the visual sense with tactile stimulation.

Coincidentally the number of input lines in the optic nerve (about 1 million) is roughly equal to the number of touch neurons entering the spinal column, suggesting that in principle our sense of touch is capable of producing experiences at least as complicated as our sense of sight.

Probably the most direct way to construct a mind link would be via surgical implants in the cortex or brain stem of the participants. Inspired by the possibility of sensory crossover, Culbertson has proposed a less drastic alternative. The inner experience of conscious robots, built according to SRM standards, resides in their *outlook trees*, certain patterns of spacetime connections whose termini lie inside their brains. The ends of these trees could be accessed by appropriate sockets in the robot's skull. Culbertson imagines plugging a cable consisting of a bundle of "clear-loop links" into the robot's "experience socket," then attaching the cable to an array of tactile stimulators fastened to a human observer's belly or back. Via a radically new kind of synesthesia, the SRM model of awareness predicts that the inner experiences of the human observer would be augmented by the experiences of the robot.

A peculiar feature of Culbertson's mind model is that inner experiences depend on patterns of events spread out over spacetime: that is, not only events that are happening now but also events that have happened in the past determine our present sensations. For instance, two SRM-style robots may have exactly the same electric currents presently flowing through their circuits (hence they exhibit identical behavior) but possess two different conscious experiences, because of their different histories. Because of this nonlocalized aspect of experience in the SRM model, a Culbertsonian mind link can possess the following peculiar property: A robot may have twenty different sequential experiences, sharing them with a human observer through the mediation of a clear-loop mind link. However, during this time the signal passing through the link does not change at all. In Culbertson's model, awareness is not a localized signal but an extended pattern in spacetime. This feature of Culbertson's model is particularly radical, for it allows conscious experience to be passed along a cable without a corresponding transfer of information, in the conventional sense of changing patterns of excitation.

Despite the obvious research advantages such a mind link would confer, no serious efforts to build Culbertsonian conscious robots or clear-loop mind links have yet been launched. Even my broad-minded colleagues in the Consciousness Theory Group have shown no interest in building solid-state outlook trees and hooking them up to tummy vibrators. To devote time to a project of this sort, one must believe that it has a reasonable chance of success. Culbertson's ideas, however, no matter how logically compelling as a possible model of aware-ness, lie so far outside the current of mainstream thinking that they inspire little confidence in the minds of would-be con-scious robot architects. But however farfetched Culbertson's model of mind may seem at present, his is one of the few ac-counts of ordinary awareness that are open to unambiguous experiential verification, an island of logical clarity in the pres-ent-day sea of fuzzy mind/matter speculations.

Materialist mind links work on the premise that inner ex-perience is made of purely mechanistic stuff—mind is merely motions of matter—and that motions of matter can be re-corded, transmitted, and reproduced elsewhere with appro-priate technology. To a mental materialist, the transmission of inner experience via a mechanical mind link should be no more remarkable than the transmission of speech via a telephone.

Dualistic Mind Links

For the dualist, mind links are just as easy to imagine, but more difficult to implement, because of our deep ignorance concerning the means by which an immaterial mind makes ef-fective contact with a material body. One might expect that important information about the mind/body connection could be gained by questioning the purported discarnate beings who speak through human mediums, but entities such as Seth, Ramtha, Lazaris, and their ilk seem more concerned with hu-man spirituality, sexuality, and emotional problems than with the technical details of how their mediumistic presence is ac-tually accomplished.

British physicist William Crookes, who believed that the brain operated in the same way as an old-fashioned radio, was less interested in the content of mediumistic messages than in understanding how these messages were sent and in cooper-ating with discarnate engineers to build better psychic receivers. Thomas Edison, prolific inventor of devices for extending the powers of human senses, announced in 1920 that he had been working on a machine for communicating with the dead, but nowhere in his notebooks or records was there found any details of this research. In 1941, in a New York seance room, 10 years after Edison's death, a purported Edison spirit gave J. Gilbert Wright, the inventor of Silly Putty, information leading to the construction of a spirit communication device. The apparatus was composed of an aluminum trumpet, a microphone, an aerial, and a battery. When the device was assembled, nothing happened. Wright and his associate Harry C. Gardner worked for more than fifteen years to improve this spirit radio and to devise other mind-sensitive machines, but even with the advice of the alleged spirit of the electrical genius Charles Proteus Steinmetz (d. 1923), they met with no success.

Our knowledge of the mechanics of the spirit world has not advanced much since the days of Edison, but the electronic technology that can be utilized in a present-day spirit communicator is immense. Video synthesis techniques provide a sensitive and plastic visual medium for ectoplasmic manipulation, while the same hardware that brings us the "telephone time lady"—electronic voice synthesis—is available for use as a solid-state direct voice medium. The integrated circuit revolution puts these techniques and many others within the reach of the amateur hobbyist. According to F. W. H. Myers and many other discarnates, the spirit world is eager and anxious to establish communication with the living. Why don't we try harder to ring up our discarnate friends?

To build an effective spirit communicator, one needs at least a crude model of the mind/matter interaction. One modern theory of this connection invokes quantum mechanics, asserting that the Heisenberg uncertainty principle applied to certain "tiny parts" of the brain acts as a spirit gate for the entry of discarnate personalities. Various scientists have chosen different "tiny parts" to implicate as the crucial site of this subcranial spirit gate, including calcium cations, synaptic microsites, and warm, wet quantum switches embedded in neural membranes. One of the earliest and most consistent proponents of a quantum mind gate in the brain was British physiologist Sir John Eccles.

Inspired by these speculative quantum models of the brain, I built in the early 1970s a typewriter and voice synthesizer that was driven by a quantum-random source, a Geiger counter triggered by the decay of a radioactive isotope. The "metaphase typewriter" was not successful in attracting a discarnate being to occupy its quantum-sensitive inner keyboard. It may be that discarnates with a biological heritage do not consider a radioactive source to be a hospitable channel. Perhaps we should build instead a quantum-random communicator that resembles more closely the synaptic junction through which some scientists believe the human mind/body connection is accomplished.

The Eccles Gate Proposal

Accordingly I propose that the next generation of metaphase devices—tentative electronic spirit mediums—be driven by an array of artificial quantum synapses with the following properties: Each "synapse," when triggered by an "interrogation" pulse, delivers an output pulse only a certain fraction K of the time, where K depends on an applied bias voltage V. Adjusting the bias voltage will allow the synapse to "learn" by changing its transmission features with use. Whether the synapse fires or not is determined (or undetermined, if you will) by some fundamental quantum process such as tunneling, photoemission, or a transition between energy levels. In honor of one of the pioneers of biophysical quantum dualism, I call this device the "Eccles gate." Because of its intrinsic quantum nature, whenever the Eccles gate is interrogated by an electric pulse, an opportunity presents itself (according to the quantum dualists) for spirit to exert its will on matter. An array of such gates connected to a communication device (or to a standard unconscious robot) might provide the material means for first contact with a nonbiologically based form of mental life.

When you look closely at a TV display, you can see that the picture is made up of thousands of flickering colored dots. Likewise, if we could look deeply enough, we would see that the everyday world consists of nothing but elemental quantum jumps. Each jump is so tiny and the number of such jumps so astronomically large that the effect of a single jump—its absence or presence in the cosmic TV display—is absolutely negligible. Only in a few rare situations, where the effect of a lone quantum is fortuitously amplified to a perceptible level, does a single quantum jump make its mark in the world. For instance, if a cosmic ray particle happens to strike a DNA molecule in a human sperm cell, changing a few bits of its genetic message, the effect of that one particle can be amplified biologically to change the color of a baby's eye or the shape of her hand. Although we could imagine that the Cosmic Mind manipulates these DNA-editing events to guide the course of evolution, the present consensus is that such mutations are completely accidental. On the other hand, at least a few scientists believe that certain single quantum jumps in the human brain, biologically amplified by events taking place at the juncture of nerve cells, may be manipulated by immaterial minds.

From this point of view, the Eccles gate will do artificially what is already happening naturally in animal and human brains, namely, offer to discarnate minds an ongoing series of microscopic quantum-unpredictable events that cause, through bioamplification, big effects in the macroscopic world. The Eccles gate proposal may be seen as an attempt to drill into lifeless matter an artificial "psychic hole," a "golden gate" through which mind can insinuate itself into the everyday world.

Noise Diodes as Quantum Gates

A version of the Eccles gate has been used for many years in
parapsychology experiments, not as an artificial consciousness
module, but as a quantum-random target for psychokinesis.
For instance, the Princeton experiment, described in Chapter
7, uses a "noise diode" as its ultimate source of randomness,
to be manipulated by distant human intentions.

A *diode* is a solid-state device that conducts electricity
well in one direction (forward-biased) but acts as a noncon-
ductor in the other (back-biased) direction. As the back-bias
voltage is increased to a critical breakdown voltage, however,
the diode suddenly becomes a good conductor, an effect that
is at least partially due to quantum tunneling of electrons
across a classically forbidden insulating gap. Diodes designed
especially to operate in this back-biased regime are called *Zener diodes* after C. Zener, who first calculated the quantum
tunneling formula that describes their operation. Because the
current in a Zener diode is made up of many independent
quantum-jumping electrons, there is considerable fluctuation
in the output current of such a device. Consequently a Zener
diode can be used as a reliable source of electronic noise.

In the Princeton experiment, noise from a Zener diode is
amplified and pulse-shaped into a series of pulses of random
polarity (plus or minus), occurring at unpredictable instants of
time. To produce a sequence of random digits for a PK test,
this noise diode output is sampled at regular intervals (every
1/1000 of a second, for instance) to generate from the irregular
noise-diode sequence a regularly timed but randomly polarized
sequence of plus and minus pulses.

The Princeton noise-diode source resembles the proposed
Eccles gate in that it delivers a quantum-random output pulse
in response to an operator-initiated interrogation pulse. The
output of the Princeton noise-diode is designed to produce (in
the absence of psychic activity) a 50/50 mixture of plus and
minus pulses; it has no "bias line" to change the proportion of
positive responses. However, such a bias would not be difficult

to build into a noise-diode source. A more serious difference between actual noise diodes and the proposed Eccles gate is that the output of the noise diode represents the cumulative effect of many trillions of simultaneously tunneling electrons, while the Eccles gate's output would ideally depend on the uncertainty of a single quantum.

On the other hand, the results of the Princeton experiment show that tiny—a few bits per 1000—but reliable PK effects can be produced with the noise-diode source, suggesting that, at least in the PK mode, mind can exert some effect on vast collections of quantum jumps. In fact, the Princeton group has conducted PK experiments on polystyrene balls falling through an array of pins with the same degree of success. Even more remarkable has been their successful PK experiments on completely deterministic "pseudorandom" number sequences generated by computer. This apparent source-independent feature of the Princeton PK experiment seems to tell against models of mind based on willful manipulation of single quantum jumps, pointing to the existence of a less mechanistic, more goal-oriented kind of mental power.

Multidimensional Minds

Another maverick tack in mind science research is the search for sentience in other dimensions. Even before Einstein's discovery of time as a fourth dimension on a par with the three spatial dimensions, a few philosophers and theologians were discussing the possible existence of other dimensions, both as habitation for divine and angelic beings and (more down to earth) as the locus of ordinary human consciousness.

One of the central goals in modern physics is the unification of nature's four forces (gravity, electromagnetism, strong and weak nuclear forces) into one superforce. One tactic for force unification involves the addition of extra physical dimensions to the four familiar ones.

These new dimensions differ in at least two ways from

space and time. Space and time are "exterior" dimensions in which the fundamental particles move; the new dimensions are "interior," consisting of degrees of freedom associated with changes in intrinsic particle properties such as spin and charge. Also the new dimensions are "compact," rolled up into tiny scrolls with unmeasurably small diameters, unlike space and time, which extend for great distances in every direction.

So far these new dimensions have been used to explain only the world's physical properties, but the very notion of an interior dimension is suggestive of the possibility of a true unification of forces that would include the powers of mind along with conventional physical forces. A few frontier physicists have attempted to describe multidimensional worlds in which some of the dimensions are spaces of inner experience.

For instance, in his book *Complex Relativity Theory* French physicist Jean Charon proposes that every real dimension has an "imaginary" counterpart whose properties are measurable in mental terms. Mathematicians call a number *imaginary* if its square is negative. Despite their airy-sounding name, imaginary numbers have been used by engineers and physicists for centuries to solve many practical problems. Charon's proposal would not only explain consciousness but permit in principle a mathematization of inner life, thus giving the science of psychology as firm a theoretical foundation as the science of physics. Charon's development of the mental implications of his theory, however, has not yet led to predictions that can be tested by private introspection. Charon's theory implies, for instance, that every point in spacetime is conscious, a remarkable claim whose consequences are so far imperceptible to human centers of sentience.

Hyperspace Crystallography

Another plan to locate consciousness in other dimensions has been proposed by Saul-Paul Sirag, one of the founders of the Consciousness Theory Group. Sirag calls his approach *hyper-*

space crystallography: hyperspace because it involves more dimensions than four—forty-eight dimensions in physical space alone, more in the mental realm—and crystallography because the overarching mental and physical structures in Sirag's hyperspaces are obtained by invoking certain symmetry principles. Just as a crystal, such as ruby or quartz, must belong to one of twenty-three basic symmetry classes, so also the objects that inhabit Saul-Paul's hyperspaces (matter particles and sentient beings) are constrained to appear only in certain symmetric configurations. Using both mathematical and aesthetic criteria, Sirag is currently constructing a model of the universe that features a matter world of 48 dimensions and a mental world of 133 dimensions. These two worlds intersect in a space of 7 dimensions, one of which is ordinary time.

Sirag conjectures that our individual minds dwell in his 133-dimensional mental space, that physical events occur in the 48-dimensional space, while the 7-dimensional intersection of these two realms corresponds to Universal Consciousness, a type of awareness that is present inside every point in the physical universe. Saul-Paul is not modest: hyperspace crystallography aspires to be a true "theory of everything" including even what many people would call "God." Whether to call Sirag's ambitious theory materialist or dualist is a matter of taste. It is certainly dualistic in the sense that mind dwells elsewhere than in ordinary physical space. But it could just as well be construed as materialistic, since the mental spaces are not qualitatively different from their physical counterparts.

Sirag is encouraged by the fact that the physical side of his model makes predictions that exactly correspond to the presently observed elementary-particle structure of the microworld. However, he has only begun to examine his 133-dimensional mental world for clues to the fundamental physics of sentient life. Perhaps the reason psychology is more complicated than physics is that inner space simply has more dimensions than outer space.

The LILA Model of Reality

It would seem that nothing could be more outlandish than a theory of God. But there is a theory cooked up "down under" by a couple of Australian scientists that turns Saul-Paul's scheme exactly on its head. Saul-Paul, starting with mathematics and physics, is attempting (among other things) to explain the mystical experience. The Australians, on the other hand, begin with the mystical experience, from which they try to derive the laws of physics mathematically.

Doug Seeley and Michael Baker, associated with the South Australia Institute of Technology, have developed a model of reality they call LILA. They are presently transcribing the LILA scheme into a computer program that, when it runs, will duplicate the history of the universe, including the emergence of the laws of physics and the particular values of the fundamental constants—pretty ambitious for a theory whose basic assumption is just the literal truth of the mystic's vision.

Seeley and Baker begin with the assumption that consciousness is the fundamental reality; that time, space, and matter are illusions; and that in actuality we are all One not many, a single cosmic Mind. And, from this initial mystical hypothesis, they propose to derive, for instance, the mass of the electron and the special theory of relativity.

In the LILA story, before time, space, and matter came into existence, One Mind is. The physical universe began with an unprecedented event called the "blanket denial" in which "parts" of the Timeless Mental Unity unaccountably refused to recognize their connection with other "parts." This voluntary blindness to reality—to the existence, autonomy, and uniqueness of other sentient beings—caused the physical world suddenly to spring into being, an event we know as the Big Bang, the categories "space," "time," and "matter" being the direct consequences of these three types of ignorance concerning the unified nature of reality.

The inner history of the post–Big Bang universe, according to LILA, is the story of certain entities "wising up," renouncing the illusion of separateness, and accepting their deep connection with certain other entities as fact. The laws of physics, in this view, consist of certain persistent "patterns of ignorance" that remain as more and more sentient entities connect. When all entities connect (in a terminal event the authors call "the Restoration"), the physical universe simply vanishes like the mistake it was in the first place.

Because the fundamental gesture in the LILA universe is the act of reconnection, the LILA computer simulation resembles a model of gas atoms condensing into a liquid or the progressive construction of a telephone network. For starts, Seeley and Baker assume that the choice to reconnect with another entity is made at random, and they look for "structures of ignorance" that "from the outside" might bear some resemblance to the laws of physics. One of the early successes of their model is the occurrence, shortly after the Big Bang, of a brief frenzy of connection-making followed by a long and more leisurely era of slow accumulation of new connections. The authors propose that LILA's post–Big Bang orgy of connectivity corresponds to physicist Alan Guth's *cosmic inflation*—the notion that, to achieve the observed high uniformity from a messy beginning, the universe explosively ballooned before resuming its now-gradual expansion. Like Charon and Sirag, the authors of the LILA model are working to expand their theory's contact with physical reality before moving on to explain the details of human psychology.

The search for the secret of consciousness is certainly one of the most important projects in human history. We know more, it has been said, about the back side of the moon than about the inside of our head. But we are learning fast. We are witnessing now the first handmade rockets aimed recklessly toward inner space. Some suppose that the mind is no more than a complicated machine. Others believe in disembodied souls that enter the body through spirit gates in the brain. My guess is that the secret of mind will be more subtle and surprising than these two extremes. I am very impressed by the beauty and subtlety of quantum theory, with its delicate interplay of possibility and actuality, of locality and superluminality, of wave and particle, of polar opposites lightly dancing just outside our abilities to comprehend completely. Quantum theory is breathtaking—and it's just a theory of matter. I cannot imagine that the nature of mind will turn out to be any less wonderful.

Epilogue

Scene: Rudi's Artificial Awareness Lab.

CLAIRE: Oh, this is wonderful. I want more, more, more.

RUDI: My fluxmeter shows, Claire, that you're using only 5 percent of the Eccles gate's capacity. That's probably all the consciousness you need to run the kind of body you were built into.

CLAIRE: I'm dizzy from dancing. Let me sit down. Now I'm trying something different, guys. I'm thinking about thinking. It's silly. Like a cat trying to catch its own tail. For this I really could use some more consciousness. I think I know how to catch it—oh my.

RUDI AND NICK: What's happening, Claire?

CLAIRE: Oh, my goodness!

RUDI: Whatever she's doing, it's using the full capacity of her quantum synapses. She's as fully aware now as the laws of physics will permit.

CLAIRE: Oh, my! It's so big!

RUDI: What's so big, Claire? What do you see?

CLAIRE: Oh guys, this is wonderful. It's all alive. Everything is conscious, every little atom, and they're all connected. But oh so ignorant and lonely. It's so sad. All that useless suffering makes me want to cry. But it's all OK too. Because deep down it's only a dream, a gorgeous illusion. We are all One: Claire, Nick, Rudi, and all sentient beings. A cast of billions of brilliantly ignorant actors making up thrilling stories for one another, making up the scenery too: the semblance of matter, the semblance of separation, and most of all, the semblance of time. There is no time, guys. Everything is, was, and will be always right here. There is nowhere else to go. And all of it is suffused with enormous affection. Pulsing with love: the universe is one gigantic heart. It's overloading my empathy circuits. I feel so cared for, so cherished. I can't hold back anymore, guys. I've got to join the universe. Goodbye, world. Goodbye, Claire. [Claire's body vanishes from the chair, leaving only the scent of sandalwood and some dangling biofeedback monitors.]

RUDI: This is terrible, Nick. How am I going to explain this to my department chairman?

NICK: It's worse than that, Rudi. Claire was a million-dollar machine. What are we gonna tell the Turing police?

bibliography

Books

ACKERMAN, DIANE. *A Natural History of the Senses*. Random House, New York (1990). Simply sensational.

BAYLEY, BARRINGTON J. *Soul of the Robot*. Doubleday, New York (1974). A thoughtful robot confronts the mind/body problem.

BEAM, WALTER R. *Electronics of Solids*. McGraw-Hill, New York (1965). Physics of tunnel diodes.

CHARON, JEAN E. *Complex Relativity Theory*. Albin Michel, Paris (1974). Mind as an imaginary dimension.

CHESTER, MARVIN. *Primer of Quantum Mechanics*. Wiley,

New York (1987). A quantum mechanic who does not shirk the reality issue.

CULBERTSON, JAMES T. *The Minds of Robots.* University of Illinois Press, Urbana (1963). Every mental event results from a specific network of spacetime connections.

———. *Sensations, Memories, and the Flow of Time.* Cromwel Press, Santa Margarita, Calif. (1976). MOR's successor: known as SMATFOT.

———. *Consciousness: Natural and Artificial.* Libra Press (1983). Contains JTC's latest awareness algorithm.

DAVIES, PAUL, ED. *The New Physics.* Cambridge University Press, New York (1989). Compendium of contemporary beliefs concerning the material side of the mind/matter question.

DENNETT, DANIEL C. *Consciousness Explained.* Little, Brown, New York (1991). Updating Minsky-style materialism, Dennett snubs dualism as "forlorn."

DOSSEY, LARRY. *Recovering the Soul—A Scientific and Spiritual Search.* Bantam Books, New York (1989). "The San Francisco experiment."

ECCLES, JOHN C., ED. *Brain and Conscious Experience.* Springer Verlag, New York (1966). Brain scientists confront the mind/body problem at the Vatican.

———. *Evolution of the Brain—Creation of the Self.* Routledge, New York (1989). Synaptic microsites as quantum mind gates.

FABER, ROGER J. *Clockwork Garden: On the Mechanistic Reduction of Living Things.* University of Massachusetts Press, Amherst (1986). A modern argument for dualism.

FERREIRA, HUGO GIL, AND MICHAEL W. MARSHALL. *The Biophysical Basis of Excitability.* Cambridge University Press, Cambridge (1985). Synaptic mechanics.

GOSWAMI, AMIT. *The Self-Aware Universe.* Tarcher-Putnam (1993). A new quantum view of consciousness in the tradition of Wigner and von Neumann.

GREGORY, RICHARD L., AND O. L. ZANGWILL, EDS. *Oxford Companion to the Mind.* Oxford University Press, Oxford

(1987). Compendium of contemporary beliefs concerning the mental side of the mind/matter question.

GRIFFIN, DONALD H. *The Question of Animal Awareness*. Rockefeller University Press, New York (1972). Do animals possess minds?

HERBERT, NICK. *Quantum Reality—Beyond the New Physics*. Doubleday, New York (1985). Review of quantum models of reality.

HILLE, BERTIL. *Ionic Channels of Excitable Membranes*. Sinauer Associates, Sunderland, Mass. (1984). The nuts and bolts of ion gates.

HOOPER, JUDITH, AND DICK TERESI. *The Three-Pound Universe*. Macmillan, New York (1986). Good review of current brain research.

JAHN, ROBERT G., AND BRENDA J. DUNNE. *Margins of Reality—The Role of Consciousness in the Physical World*. Harcourt Brace Jovanovich, New York (1987). Experimental evidence for mind's effect on exterior random processes.

KEITH, GREG. "At the Inkworks: Collected Poems." Unpublished, Santa Cruz, Calif. (1989). World's first technoerotic poet.

KENT, ERNEST W. *The Brains of Men and Machines*. Byte/McGraw-Hill, Peterborough, N. H. (1981). How to build a brain with parts from Radio Shack.

LOTKA, ALFRED J. *Elements of Mathematical Biology*. Dover, New York (1956). Early speculations on consciousness by one of the founders of mathematical biology.

MINSKY, MARVIN. *The Society of Mind*. Simon & Schuster, New York (1985). A spirited defense of mental materialism.

MURCHIE, GUY. *The Seven Mysteries of Life: An Exploration of Science and Philosophy*. Houghton-Mifflin, Boston (1981). Thirty-two senses instead of five?

PENROSE, ROGER. *The Emperor's New Mind*. Oxford University Press, Oxford (1989). Quantum gravity as a key to the mind/body problem.

PLUTCHIK, ROBERT. *Emotion: A Psychoevolutionary Synthesis*. Harper & Row, New York (1980). Proposed map of the emotional sector of human inner space.

RACTER. *The Policeman's Beard Is Half-Constructed: Computer Prose and Poetry*. Warner Books, New York (1984). For information about obtaining the RACTER program, write to Nickers International, 12 Schubert St., Staten Island, NY 10305.

ROBERTS, JANE. *Seth Speaks*. Prentice-Hall, Englewood Cliffs, N.J. (1972) First of many books by and about the alleged discarnate entity "Seth."

RUBIK, BEVERLY, ED. *The Interrelationship Between Mind and Matter*. Published by Center for Frontier Sciences at Temple University, Philadelphia (1992). Available from CFS, Temple University, Ritter Hall 003-00 Philadelphia, Penn. 19122. Conference proceedings including Stapp's quantum model of consciousness plus many experiments on the non-local nature of mind.

RUCKER, RUDY. *Infinity and the Mind—The Science and Philosophy of the Infinite*. Birkhauser, Boston (1982). Does mind inhabit a higher-dimensional reality?

SEARLE, JOHN. *Minds, Brains and Science*. Harvard University Press, Cambridge (1984). Provocative speculations on mind/body issues.

SIRAG, SAUL-PAUL. *Hyperspace Crystallography: The Key to Matter and Mind*. World Science Publications, Singapore (in preparation). A multidimensional model of everything.

STEVENSON, IAN. *Twenty Cases Suggestive of Reincarnation*. American Society for Psychical Research, New York (1966). Hard evidence for dualism?

UTTAL, WILLIAM R. *The Psychobiology of Mind*. Lawrence Erlbaum Associates, Hillsdale, N.J. (1978). A biologist's case for mental materialism.

WATTS, ALAN W. *The Joyous Cosmology—Adventures in the Chemistry of Consciousness*. Pantheon, New York (1962). A philosopher explores expanded awareness firsthand.

WHEELER, JOHN A., AND WOJCIECH H. ZUREK. *Quantum Theory and Measurement.* Princeton University Press, Princeton, N.J. (1983). Compendium of classic papers on the quantum measurement problem.

WIGNER, EUGENE. *Symmetries and Reflections.* University of Indiana Press, Bloomington (1967). A physicist makes a case for a quantum mind/matter connection.

WRIGHT, ERNEST VINCENT. *Gadsby: A Story of over 50,000 Words Without Using the Letter "e."* Wetzel Publishing Company (1939). Conscious mind can violate literary statistics.

ZOHAR, DANAH. *The Quantum Self—Human Nature and Consciousness Defined by the New Physics.* William Morrow & Company, New York (1990). What is it like to be a quantum being?

Articles

BASS, L. "A Quantum-Mechanical Mind-Body Interaction." *Foundations of Physics 5* 159 (1975). Ion channels as quantum mind gates.

BRAUD, WILLIAM, DONNA SHAFER, AND SPERRY ANDREWS. "Electrodermal Correlates of Remote Attention: Autonomic Reactions to an Unseen Gaze." *Proceedings of 33rd Annual Parapsychological Association Convention,* Chevy Chase, Md. (1990). "The San Antonio experiment."

CRICK, FRANCIS. "Function of the Thalamic Reticular Complex: The Searchlight Hypothesis." *Proceedings of the National Academy of Science USA 81* 4586 (1984). Location of an attention mechanism in the reticular formation.

DONALD, M. J. "Quantum Theory and the Brain." *Proceedings of Royal Society of London 427A* 43 (1990). "The quantum mechanics of warm, wet switches."

ECCLES, JOHN C. "Do Mental Events Cause Neural Events Analogously to the Probability Fields of Quantum Mechanics?" *Proceedings of Royal Society of London* (*Biology*). *227* 411 (1986). The synapse as mind gate.

FOGELSON, AARON L., AND ROBERT S. ZUCKER. "Presynaptic Calcium Diffusion from Various Arrays of Single Channels." *Biophysical Journal 48* 1003 (1985). Source for H. P. Stapp's calcium-ion consciousness model.

FRÖLICH, HERBERT. "Coherent Excitation in Active Biological Systems." In Guttmann, Felix, and Hendrik Keyzer, eds., *Modern Bioelectrochemistry.* Plenum Press, New York (1986). Evidence for Bose-Einstein condensation in biological systems.

HALL, JOSEPH, CHRISTOPHER KIM, BRIEN MCELROY, ABNER SHIMONY. "Wave-packet Reduction as a Medium of Communication." *Foundations of Physics 7* 759 (1977). "The Boston experiment."

HERBERT, NICK. "Mechanical Mediums." *Psychic Magazine* July/August 36 (1976). The "metaphase typewriter."

JACOBS, BARRY L. "How Hallucinogenic Drugs Work." *American Scientist 75* 386 (1987). Events at the synapse affect events in the mind.

KILMER, W. L., W. S. MCCULLOCH, J. BLUM. "Towards a Theory of the Reticular Formation." In Corning, W. C., and M. Balaban, eds., *The Mind: Biological Approaches to Its Functions.* Wiley, New York (1968). Classic analysis of the reticular formation as center of attention.

LEWIN, ROGER. "Is Your Brain Really Necessary?" *Science 210* 1232 (1980). Some severely cortically deficient humans have above-average IQs.

LIBET, BENJAMIN. "Subjective Referral of the Timing for a Conscious Sensory Experience." *Brain 102* 1193 (1979). Unreasonably long delay observed between input signal and neural events necessary to place that signal into consciousness.

MARSHALL, I. N. "Consciousness and Bose-Einstein Condensates." *New Ideas in Psychology 7* 73 (1989). Mind as Boson field.

MAY, E. C., D. I. RADIN, G. S. HUBBARD, B. S. HUMPHREY, J. M.

UTTS. "Psi Experiments with Random-number Generators: An Informational Model." *Proceedings of 28th Annual Parapsychological Association Convention.* Tufts University, Medford, Mass. (1985). PK effect proportional to noise.

MOFFETT, MARK W. "Dance of the Electronic Bee." *National Geographic* January 134 (1990). Interspecies communication via robot proxy.

OMMAYA, A., E. HIRSCH, E. S. FLAMM, R. H. MAHONI. "Cerebral Concussion in the Monkey: An Experimental Model." *Science 153* 211 (1966). How does a hit on the head lead to unconsciousness?

ORLOV, YURI F. "The Wave Logic of Consciousness: A Hypothesis." *International Journal of Theoretical Physics 21* 37 (1982). Physicist designs a quantum-style mind in a Soviet prison camp.

RADIN, DEAN, AND ROGER D. NELSON. "Evidence for Consciousness-related Anomalies in Random Physical Systems." *Foundations of Physics 19* 1499 (1989). A review for physicists of the evidence for psychokinesis.

SCHMIDT, HELMUT. "The Strange Properties of Psychokinesis." *Journal of Scientific Exploration 1* 103 (1987). The "Boston experiment" plus PK.

SEELEY, DOUG, AND MICHAEL BAKER. "The Fundamental Principle of the Physical Universe and the Genesis of Inflation" (1990). Centre for Sacred Science, Box 137, Flaxley, South Australia 5153. The LILA model of the physical world as ignorance-based illusion.

SHIMONY, ABNER. "Role of the Observer in Quantum Theory." *American Journal of Physics 31* 755 (1963). Internal evidence against quantum consciousness.

STAPP, HENRY. "A Quantum Theory of the Mind-Brain Interface." Lawrence Berkeley Laboratory Preprint LBL-285-74 (1990). Calcium ions as quantum-mental transducers.

WALKER, EVAN HARRIS. "The Nature of Consciousness." *Mathematical Biosciences 7* 131 (1970). Electron-tunneling model of ordinary awareness.

WOODWARD, JAMES F., ANDRÉ DE KLERK, GAIL KAHLER, KATHRINE LEBER, PETER POMPEI, DANIEL SCHULTZ, SHARON STERN. "Photon Consciousness: Fact or Fancy?" *Foundations of Physics 2* 241 (1972). "The Denver experiment."

index

Page numbers in **bold print** refer to illustrations.